T0091531

Information Retrieval in Bioinformatics

Soumi Dutta · Saikat Gochhait
Editors

Information Retrieval in Bioinformatics

A Practical Approach

palgrave
macmillan

Editors
Soumi Dutta
Institute of Engineering &
Management
Kolkata, West Bengal, India

Saikat Gochhait
Symbiosis Institute of Digital
and Telecom Management
Symbiosis International (Deemed
University)
Pune, Maharashtra, India

ISBN 978-981-19-6505-0 ISBN 978-981-19-6506-7 (eBook)
https://doi.org/10.1007/978-981-19-6506-7

This Palgrave Macmillan imprint is published by the registered company Springer Nature Singapore Pte Ltd.
The registered company address is: 152 Beach Road, #21-01/04 Gateway East, Singapore 189721, Singapore

Contents

Editors and Contributors

About the Editors

Dr. Soumi Dutta She did her B.Tech in Information Technology from WBUT and M.Tech in Computer Science Engineering from WBUT with securing 1st position (Gold medal). She has Ph.D. from Indian Institute of Engineering Science and Technology (IIEST, Shibpur).

Dr. Saikat Gochhait is Post-Doctoral Fellow from University of Extremadura, Spain. He has been awarded with MOFA Taiwan Fellowship and Ministry of Health Russian Fellowship. He has more than 50 publications in journals indexed in Scopus, WoS, ABDC, Google Scholar, etc. And 6 books published with Springer and IGI Global indexed in Scopus. He features in the P-Rank: A Publication Ranking, HCERES 2021.

Contributors

Dr. Azariah Babu completed Ph.D. under the guidance of Prof. Dr. T. N. Ananthakrishnan and done Post-Doctoral Research under the guidance of Professor (Mrs.) Silvia Dorn at the SWISS Federal Institute of Technology Zurich (ETH) in the Institute of Plant Sciences, Applied Entomology Zurich, Switzerland. He has more than 28 years of research experience in basic and applied aspects of Entomology. He has been engaged in various aspects of insect-host plant-natural enemy interactions during the past 28 years.

Nikhil Danny Babu is a Research Scholar in Physics at the Indian Institute of Technology Guwahati, Guwahati, India. He has a BS-MS dual degree in Physics from the Indian Institute of Science Education and Research—Thiruvananthapuram with an INSPIRE fellowship from the Department of Science and Technology (DST)—Govt. of India. His main areas of research interest lie in theoretical condensed matter physics, particularly in low-dimensional correlated electron systems. He has published two papers in leading international journals in condensed matter physics. He is also deeply interested in recent advances in AI and machine learning methods and its applications to problems in the natural sciences. He has some experience working as a teaching assistant for the undergraduate courses PH101 (Introduction to Classical Mechanics, Special Relativity and Quantum Mechanics), PH102 (Introduction to Electromagnetism), and for the PhD level course PH707 (Computational Physics) offered by the Department of Physics at IIT Guwahati. He is an avid movie buff and loves bird photography.

Dr. Chittaranjan Baruah research is related to the fields of (a) Bioinformatics: Structural biology, proteomics, molecular phylogeny; (b) Zoology: Fish Biology & Fish biotechnology, Nano-bio-pesticides; and (c) Conservation biology: Turtles & Tortoises of Northeast India.

Dr. Bhabesh Deka (Senior Member, IEEE) is currently a Professor with the Department of Electronics and Communication Engineering, Tezpur University. He leads the Computer Vision and Image Processing (CVIP) Laboratory, Department of ECE, Tezpur University. He is also a Principal Investigator of two major research projects sponsored by All India Council for Technical Education (AICTE) and Indian Space Research Organisation (ISRO), Government of India. His research interests include image processing, particularly inverse ill-posed problems, computer vision, compressive sensing MRI, and machine learning for biomedical signal/image analysis.

Dr. Uma Dutta is Associate Professor and currently Head, Department of Zoology, and acting Dean of Life Sciences, Cotton University, Guwahati. She has around 29 years of teaching experience from Sr. Secondary to Ph.D. courses in the field of Zoology esp. Cell and Molecular Biology, Computational Biology and Bioinformatics, Biochemistry,

Developmental Biology, Toxicology, Physiology, etc. She is currently a mentor of several bachelor's, master's, internees, and Ph.D. students. She has published several papers in leading international journals of repute and authored several book chapters and reviewed many research papers. She is the recipient of many prestigious awards, a life member with many scholarly societies and academic bodies, and has produced a number of Ph.D. students. Her field of research interests are a) mitigation and therapeutic approaches to cancer by natural bioactive components; b) hormonal interaction with herbal components in infertility issues; c) environmental toxicology; and d) insect pest management with biopesticides and biodiversity with reference to avian fauna.

Recently, she has diverted her attention to using machine learning methods to solve problems in the life sciences and human diseases.

Dr. Annika Durve Gupta is working with B.K Birla College (A) in the Department of Biotechnology.

P. Keerthana lives in Chennai and started her career as Electronics and Communication Engineer and currently pursuing a master's in Analytics and Finance from Symbiosis Institute of Digital and Telecom Management. As an avid reader, her dream of writing started during her college days and has published a paper on IoT-based medical record-tracking systems for patients. Her strong fascination for the Internet of Things has driven her interest in the emerging field of Translational Bioinformatics.

Vivek Kumar is associated as Research Fellow at GNA University, Jalandhar, Punjab, India.

Dr. Saurov Mahanta is associated with National Institute of Electronics and Information Technology, Guwahati with major research area on Bioinformatics.

Dr. Ritu Pasrija currently works at the Department of Biochemistry, Maharshi Dayanand University. She does research in Fungal Genetics, Drug Resistance and Organelle Contacts.

Sonali Patil is associated as Assistant Professor at the Department of Bioanalytical Sciences, B.K Birla College (Autonomous), Kalyan, Maharashtra.

Dr. D. Ramyachitra is affiliated to Department of Computer Science, Bharathiar University. He has published numerous publications in various national and international peer-reviewed journals and presented scientific papers across the world. Because of the active association with different societies and academies as well as the contributions, she has been recognized by the subject experts around the world. Her contributions are appreciated by various reputed awards. Her clinical and scientific research interests include Grid Computing, Data Mining, and Bioinformatics.

Dr. Vijay Rana is working with GNA University, Punjab in the Department of Computer Science.

Dr. M. Revathi is working with Bharathiar University in the Department of Biotechnology.

Dr. Girish S. Setlur, Ph.D. is Professor of Physics and Adjunct Professor at the Mehta Family School of Data Science and Artificial Intelligence (MFSDSAI) at the Indian Institute of Technology Guwahati, Guwahati, India. He earned his B.Tech. in Engineering Physics from the Indian Institute of Technology Bombay, India. He earned his M.S. and Ph.D. degrees in Physics from the University of Illinois at Urbana-Champaign, USA. He has more than three decades of teaching experience both at the undergraduate and postgraduate levels. He is currently a mentor of several bachelor's, master's, and Ph.D. students. He has published numerous papers in leading national and international journals and presented invited and contributed papers in various national and international conferences. His graduate-level textbook on Classical and Quantum Fields has been published by CRC Press (Taylor and Francis). He is also co-author of a first-year undergraduate Physics textbook with Prof. Eric Mazur, Harvard University called Principles and Practice of Physics (Pearson's publication). His specialized research interests include theoretical condensed matter physics and machine learning and its applications along with broad interests in all fields of science and engineering. His hobbies include listening to music and playing violin.

Dr. Sunny Sharma is working with University of Jammu in the Department of Biotechnology.

Dr. Sunita is working with Arni University, Himachal Pradesh in the Department of Computer Science.

List of Figures

Chapter 7

List of Tables

CHAPTER 1

Bioinformatics Overviews

Ritu Pasrija

1 Background

In the 1970s, a Dutch theoretical biologist Paulien Hogeweg along with Ben Hesper, first coined the term bioinformatics. They were interested in accumulating information regarding biological systems. Their observation was that in addition to biochemistry and biophysics, it is worthwhile to recognise bioinformatics as a research area and has the potential to become 'biology of the future'. This became true as in these last 50 years, development of bioinformatics has happened at a very fast pace. Although for a particular interval, persons viewed bioinformatics as the software tools advancement method to support, accumulate, manoeuvre, and scrutinise biological information. Whilst this application is indeed a significant one in bioinformatics, this field has much more potential than that. Both 'bioinformatics' and 'computational biology' are instrumental in accumulating enormous information of several parts of natural science. So, it is important to understand the difference in these two terms. On the one hand, bioinformatics uses computer science,

R. Pasrija (✉)
Department of Biochemistry, Maharshi Dayanand University, Rohtak, India
e-mail: ritupasrija.biochem@mdurohtak.ac.in

S. Dutta and S. Gochhait (eds.), *Information Retrieval in Bioinformatics*,
https://doi.org/10.1007/978-981-19-6506-7_1

statistics to molecular biology and create computational & statistical techniques, which help in examination and management of biological data. On the other hand, computational biology uses computational simulation mode, mathematical models, and fundamentals in computer science, genetics, anatomy, biochemistry, and statistics among others. Amalgamation of these two has led to a new term called 'Systems biology', which combines organism-wide information of natural science for acquiring a comprehensive perception of a biological entity, like a bacterium. This led to creating synthetic genomes, and soon a synthesised cell would become a reality. Nonetheless, to understand the feasibility of this fancy hypothesis, it is important to revisit the key discoveries in biological sciences, which would also help in understanding the history of development of 'bioinformatics' as a separate branch.

2 History

During the 1950s, DNA and computers were not the important tools in research and in biochemistry, investigations were largely happening on mechanistic enzymes model. Many scientists in fact thought that proteins are the carriers of genetic information, as DNA seemed too simple to carry genetic information, whereas protein show a large number of alternatives and complexity.

This, the major turning point in bioinformatics has to be DNA being regarded as the genetic material. The first evidence for this came from experiments of Oswald Avery et al. (1944), who revealed that DNA regulates the characters in organisms, instead of proteins. This group studied the uptake of pure DNA from a virulent *Streptococcus pneumoniae (S. pneumonia)* bacterial strain, which has smooth round colonies (named S); which could bestow virulence to even a non-virulent strain (rough colonies, R) (Avery et al., 1944). Subsequent work by Alfred Hershey and Martha Chase (in 1952) validated these findings that DNA of bacterial cells infected by bacteriophages can be transmitted to other bacterium and alter the phenotype of recipient cell (Hershey & Chase, 1952). Later in 1953, James Watson, Francis Crick, and Rosalind Elsie Franklin finally proposed the double-helix structure of DNA (Franklin & Gosling, 1953; Watson & Crick, 1953). Further, it took additional 13 years in interpreting the amino acid codon and 24 additional years in improving the first DNA sequencing technique. Thus, 1970–1980 witnessed a paradigm shift from protein to DNA analysis. In 1970, Saul B. Needleman and

Christian D. Wunsch established their dynamic programming algorithm for the alignment of pair-wise protein sequences. After another decade of first multiple sequence alignment (MSA) algorithms was developed, its application was applied to other biological sequences (DNA and RNA). Therefore, development in DNA always lagged behind proteins. Comparison wise, the amino acids alignment is based on-identical, similar and dissimilar amino acids (based on their chemical properties), whereas only identical matches and mismatches are considered for DNA and RNA. The MSA and its use to sequence-structure-function relationship is so common and useful that from the 1980s onwards, the term 'bioinformatics' is mostly used to refer to the computational methods for genomic data analysis. This big transformation opened the possibility to sequencing whole genomes. Biologist, Fred Sanger, undertook the first genome sequencing in a bacterial virus, called bacteriophage $\phi \times 174$ (5368 base pairs). Later, Craig Venter in the 1980s sequenced the first organism, a bacterium *Haemophilus influenza*. Ernst Haeckel, in 1866, used DNA sequences in phylogenetic inference and reconstructed the first molecular phylogenetic trees from amino acids linear arrangements in proteins, which show the closeness and related ness among species during evolution.

Similarly, many more genomes are already sequenced (more than 1000), including of human, called "The Human Genome Project (HGP)", which completed in April 2001 by two independent groups (McPherson et al., 2001; Venter et al., 2001). These recent life science data explosions—such as genotyping, transcriptomics, or proteomics—also became possible with the availability of genomes and opened gates for new studies. The gold mine of enormous data later became freely available at European Molecular Biology Laboratory (EMBL) site (www.ensembl.org). Along with this, various bioinformatics tools also became available including on this site, like—BLAST, Ensembl, primer synthesis, phylogeny that rely on the accessibility of the cyberspace. Investigations on genomic sequences information, including humans unlocked the applied prospects like—discovery of drug and their targets, as well as individualised therapy. Thus, biologists end up being increasingly dependent on computational scripts, written in scripting languages, such as C, C++, Shell, Python, R, and Ruby. Before, we look at the research challenges, algorithms, big data, retrieval of information, and application of bioinformatics; it is crucial to understand the basics of biology and bioinformatics (Porter et al., 2021).

Basics of Bioinformatics: We know that all surviving entities are composed up of units called cells, which contain the genetic material (nucleic acid), and passed from one generation to the next. Many living systems are made up of only one cell (unicellular), and one cell is the whole organism; whereas in developed species like plants and animals, a life form has more than billions of cells. Apart from that, cells can be of two major types: prokaryotic cells and eukaryotic cells. The prokaryotic cells do not have nucleus and genetic material lies open inside the cell, whereas in eukaryotic cells genetic material is present inside a structure, called nucleus. Prokaryotic cells are generally unicellular, whereas eukaryotic cells can be either unicellular, like baker's yeast *Saccharomyces cerevisiae* (*S.cerevisiae*) or multicellular organisms, like humans.

The nucleic acid is of two kinds in nature: deoxyribonucleic acid (DNA) and ribonucleic acid (RNA). Most organisms' genetic material is DNA, although few cells might have RNA, as in certain viruses. Both forms of nucleic acid are polymers, assembly of repeated units, called nucleotides. A nucleotide involves three segments: a base, a pentose sugar (ribose), besides a phosphate group. These bases vary in different nucleotides and are of four types: Guanine (G), Thymine (T), Adenine (A) and Cytosine (C). For RNA, the bases are identical like DNA, except T replaced with Uracil (U). RNA is a single helix, whereas DNA is a double-helix molecule, in which bases lie parallel to each other and form bond with each other, called hydrogen bond base pairing. During bond formation, T always pairs with A base through two hydrogen bonds (double bond) and C always pairs with G through three bonds (triple bond). This removes steric hindrance and stabilises the structure. When an RNA strand duo with another strand, the pairing followed is $A = U$ and $C \equiv G$. These nucleotides join one after another linearly and generate polymer.

***The Central Dogma*:** DNA is the controlling element, but needs to pass on the information to a kind of RNA, termed as messenger RNA (mRNA), through a process called transcription and subsequently data in mRNA is passed via translation to proteins. Three DNA/RNA nucleotides code for one amino acid, which polymerise in a linear fashion to make proteins. Like UUU code for an amino acid, phenylalanine and AUG code for methionine. There are 20 standard amino acids, which are written as alphabets in capital. The sequence of three DNA nucleotides (called codon) decides the amino acid incorporated. The nucleotides (U,

C, A and G) in random sequence of three can give 64 possible combinations, thus one amino acid may be coded by more than one combination and grouped together. Among these 64, three codons act as stop codon (UAA, UGA and UAG), and once incorporated, they stop the translation, as they do not code for any amino acid.

The genetic material has different genes on it, which control the different characters in an organism, its fitness, etc. Thus, scientist developed fantasy for this molecule, as genetic material manipulation is possible, and desired characters, fitness, chances of survival is achievable. Similarly, it also opened avenues for gene manipulation to correct any disease phenotype. All this became possible, as research has given insight into function of the genes. This led to immense advancement of techniques to study genetic material, termed 'genomics'. Interestingly, the different cells of an organism although have same DNA, but do not express all the genes present on their genetic material. Thus, all cells do express some common genes, called 'house-keeping' genes, whereas some genes are transcribed to RNA and to translated proteins in exclusive cells or time interval. Like in foetal stage, the haemoglobin synthesised is different from that of an adult and are product of different genes. Once born, the foetal haemoglobin gene is not transcribed any more. These differential DNA expressions pattern led to different RNA profile in various cells and is studied under 'transcriptomics'. Similarly, the functional molecules, proteins presence, and their measure are covered in 'proteomics'. Apart from that, three-dimensional (3D) modelling of biomolecules and biological systems is also of interest to biological scientist.

All these advancements led to generation of a large amount of biological information among different organisms and species. Thus, biological scientists need to utilise computational and analysis tools for acquiring, understanding, interpreting and sharing of these data. These have led to the 'Bioinformatics' we know today, and became essential in organising information in modern biology, as well as treatment. Therefore, it is not exaggerating to conclude that bioinformatics is a multidisciplinary field, which connects biology with computer's knowledge, mathematics, statistics, and physics. This has led scientists in essentially acquiring a good knowledge of molecular biology, along with computer science for analysis of bioinformatics data.

The Human Genome: The name 'genome' exactly refers to total genes or whole of the DNA's content in an organism or a cell. A regular human cell encloses 23 couples of chromosomes (separate threads of DNA). The human genome has 22 autosomes, and female additionally has two copies of X chromosome (XX) and male has one X and one Y chromosome (XY). These 23 pairs of DNA threads have approximately ~ 20,000–25,000 genes in the human genome, which are generally protein coding, although sometimes only RNA is transcribed, which has regulatory role. Besides protein coding region, the controlling sequences are also there in genome, which includes—promoters, introns, intergenic (between-gene) regions, and repetitive sequences in the genome. On an average, more than half of the human genome is transcribed and translated, though a very small quantity of them is managed as mRNAs and study of all RNA transcripts is included under 'transcriptomics'.

3 Perspectives of Computer Science and Information Technology

As earlier stated, the sequencing techniques opened an era of loads of data, called big data and its usefulness relies on programming and software development, and building enormous datasets of biological information in research (Gochhait et al., 2021). This information is much more than easily and efficiently interpreted by a biology researcher. Further, different investigators may decipher the data in dissimilar ways, and even the same researcher may make varying explanations, resulting in erratic and non-uniform data processing. Sometimes, after interpretation, the rationale behind may be lost or only loosely remembered. Occasionally, a researcher exits a study group; the technique used to understand data also goes with them and vanishes. Lastly, the researcher's interpretation may be prejudiced towards getting a predetermined consequence. Thus, bioinformatics being a systematic study rule out all these anomalies in science and research. The important basis of bioinformatics includes.

Algorithms: These are the rules followed in computations and done on both Linux and Windows platform. Various tasks rely on particular algorithms, which are critical in both examining and accurately handling of the data. The biological scientists and bioinformaticians choose the computer science processes for sequencing, gathering, unravelling biological functions and relationships, which finally helps in interpretation of the information. These include DNA, RNA and protein alignments (could be local and global), gene prediction, phylogenetic tree construction database similarity search, motif detection, Markov chains or information entropy and sequence logos, molecular modelling, etc. The FASTA, BLAST, Spectral Forecast, Objective Digital Stains (ODSs), self-sequence alignment and Discrete Probability Detector (DPD) algorithm are few examples. A bioinformatics can use resources freely available over the internet and some are paid softwares. The literature references and various databases (genome, sequence, function structure) of molecular biology can be searched at PubMed, PubMed Central, NCBI, EBI, ExPASy, RSCB. Some important tools in Bioinformatics are introduced below:

1. **Recovery and exploration of sequence**: Linear sequence of DNA, RNA and proteins is used for sequence alignment and provides lots of information. It depends on particular algorithm like FASTA BLAST, CLUSTAL X/W, etc. These are helpful for identity, similarity and dissimilarity in sequences (homology), phylogeny search analysis and phylogeny tree construction for relatedness among species.

FASTA is a text-based arrangement for representing either base sequences or amino acid (protein) order, in which base or amino acids are symbolised with single-alphabet codes and blanks in between alphabets are not permitted. Matches are displayed in black and red are treated as mismatches in nucleotide alignment. It was developed by David J. Lipman and William R. Pearson in 1985 and is used in many programming languages like Python, PERL, Ruby, etc. A multiple sequence FASTA layout is obtainable by concatenating various single FASTA sequence in a single file. The FASTA sequence first line starts with a '>' (greater-than) symbol as shown below for two genes: Green fluorescent protein (GFP) from jellyfish and insulin from humans.

GFP gene sequence of *Aequorea victoria* in FASTA format

```
>AGTAAAGGAGAAGAACTTTTCACTGGAGTTGTGACAATTCTTGTTGAATTAGATGGTGAT
GTTAATGGTCACAAATTTTCTGTTAGTGGAGAGGGTGAAGGTGATGCAACATACGGAAAAC
TTACCCTTAAATTTATTTGTACTACTGGAAAACTACCTGTTCCCTGGCCAACACTTGTTAC
TACTTTGACTTATGGTGTTCAATGTTTTTCAAGATACCCAGATCACATGAAACGGCACGAC
TTTTTCAAGAGTGCAATGCCCGAAGGTTATGTACAAGAAAGAACTATTTTTTTCAAAGATG
ACGGTAACTACAAGACACGTGCTGAAGTTAAGTTTGAAGGTGATACCCTTGTTAATAGAAT
CGAGTTAAAAGGTATTGATTTTAAAGAAGATGGAAACATTCTTGGACACAAATTGGAATAC
AACTATAACTCACACAATGTATACATTATGGCAGACAAACAAAAGAATGGAATCAAAGTTA
```

```
ACTTCAAAATTAGACACAACATTGAAGATGGAAGTGTTCAACTAGCAGACCATTATCAACA
AAATACTCCAATTGGCGATGGCCCTGTTCTTTTACCAGACAACCATTACCTGTCCACACAA
TCTGCTCTTTCTAAAGATCCCAACGAAAAGAGAGACCATATGGTGCTTCTTGAGTTTGTAA
CAGCTGCTGGTATTACACACGGTATGGATGAACTATACAAACACCATCACCATCACCATCA
CTAG
```

Humans Insulin gene sequence in FASTA format

```
>AGCCCTCCAGGACAGGCTGCATCAGAAGAGGCCATCAAGCAGGTCTGTTCCAAGGGCCT
TTGCGTCAGGTGGGCTCAGGATTCCAGGGTGGCTGGACCCCAGGCCCCAGCTCTGCAGCAGG
GAGGACGTGGCTGGGCTCGTGAAGCATGTGGGGGTGAGCCCAGGGGCCCCAAGGCAGGGCACC
TGGCCTTCAGCCTGCCTCAGCCCTGCCTGTCTCCCAGATCACTGTCCTTCTGCCATGGCCCTG
TGGATGCGCCTCCTGCCCCTGCTGGCGCTGCTGGCCCTCTGGGGACCTGACCCAGCCGCAGCC
TTTGTGAACCAACACCTGTGCGGCTCACACCTGGTGGAAGCTCTCTACCTAGTGTGCGGGGAA
CGAGGCTTCTTCTACACACCCAAGACCCGCCGGGAGGCAGAGGACCTGCAGGGTGAGCCAACT
GCCCATTGCTGCCCCTGGCCGCCCCCAGCCACCCCCTGCTCCTGGCGCTCCCACCCAGCATGG
GCAGAAGGGGGCAGGAGGCTGCCACCCAGCAGGGGGTCAGGTGCACTTTTTTAAAAAGAAGTT
CTCTTGGTCACGTCCTAAAAGTGACCAGCTCCCTGTGGCCCAGTCAGAATCTCAGCCTGAGGA
CGGTGTTGGCTTCGGCAGCCCCGAGATACATCAGAGGGTGGGCACGCTCCTCCCTCCACTCGC
CCCTCAAACAAATGCCCCGCAGCCCATTTCTCCACCCTCATTTGATGACCGCAGATTCAAGTG
TTTTGTTAAGTAAAGTCCTGGGTGACCTGGGGTCACAGGGTGCCCCACGCTGCCTGCCTCTGG
GCGAACACCCCATCACGCCCGGAGGAGGGCGTGGCTGCCTGCCTGAGTGGGCCAGACCCCTGT
CGCCAGGCCTCACGGCAGCTCCATAGTCAGGAGATGGGGAAGATGCTGGGGACAGGCCCTGGG
GAGAAGTACTGGGATCACCTGTTCAGGCTCCCACTGTGACCTGCCCCGGGGCGGGGGAAGGAG
GTGG
GACATGTGGGCGTTGGGGCCTGTAGGTCCACACCCAGTGTGGGTGACCCTCCCTCTAACCTGG
GTCCAGCCCGGCTGGAGATGGGTGGGAGTGCGACCTAGGGCTGGCGGGCAGGCGGGCACTGTG
TCTCCCTGACTGTGTCCTCCTGTGTCCCTCTGCCTCGCCGCTGTTCCGGAACCTGCTCTGCGC
GGCACGTCCTGGCAGTGGGGCAGGTGGAGCTGGGCGGGGGCCCTGGTGCAGGCAGCCTGCAGC
CCTTGGCCCTGGAGGGGTCCCTGCAGAAGCGTGGCATTGTGGAACAATGCTGTACCAGCATCT
GCTCCCTCTACCAGCTGGAGAACTACTGCAACTAGACGCAGCCCGCAGGCAGCCCCACACCC
GCCGCCTCCTGCACCGAGAGAGATGGAATAAAGCCCTTGAACCAGC
```

Similarly, protein sequence in FASTA format for both proteins can be written and provided below—

GFP protein sequence from *Aequorea victoria* in FASTA format

```
>MSKGEELFTGVVPILVELDGDVNGHKFSVSGEGEGDATYGKLTLKFICTTGKLPVPWPTLVT
TFSYGVQCFSRYPDHMKQHDFFKSAMPEGYVQERTIFFKDDGNYKTRAEVKFEGDTLVNRIEL
```

```
KGIDFKEDGNILGHKLEYNYNSHNVYIMADKQKNGIKVNFKIRHNIEDGSVQLADHYQQNTPI
GDGPVLLPDNHYLSTQSALSKDPNEKRDHMVLLEFVTAAGITHGMDELYK
```

Insulin protein in Humans in FASTA format

```
>MALWMRLLPLLALLALWGPDPAAAFVNQHLCGSHLVEALYLVCGERGFFYTPKTRREAED
LQVGQVELGGGPGAGSLQPLALEGSLQKRGIVEQCCTSICSLYQLENYCN
```

BLAST: Its full form is **B**asic **L**ocal **A**lignment **S**earch **T**ool and this procedure catches the region of similarity between sequences. This was first proposed in 1990 by David J. Lipman and his team, and one of highly cited paper with more than 65,000 citations (Altschul et al., 1990). This program can compare both nucleotide (n-BLAST) and protein primary sequences (p-BLAST) in two different variants. The evaluation of sequence is finally used to calculate the statistical significance of matches. It is freely available on internet at 'https://blast.ncbi.nlm.nih.gov/Blast.cgi'. It can be performed on various operating systems like UNIX, Linux, Mac, and MS Windows and is written in C and C+ language. BLAST deduce useful and evolutionary relationships between linear arrangements of bases/amino acids, as well as help identify members of gene families. Like haemoglobin gene sequence in humans can be compared with that of mouse. The input sequence is generally provided in FASTA or gene bank format; whereas the output formats, include HTML, plain text, and XML. Besides online, the program is also available in free and paid download versions (BLAST+). The megablast and discontiguous megablast are other variants with separate applications.

Clustal W/X: is an algorithm for multiple sequence analysis (MSA) of DNA or proteins. It produces meaningful multiple sequence alignment of divergent species, and computes best matches for the chosen sequences and line them up so that the resemblances and the disparities can be understood by viewing the cladograms and phylograms, both of which are part of the Clustal W algorithm.

2. **Graph theory:** Graph theory, also called as graph methodology, takes the help of graphs for comparison. Graph is a set of vertices (lines), connected by edges (exist between two vertices), which could be directed (with arrow), undirected, weighed, etc. It is useful in showing networks or flow of communication and simplifies complex relationship.

3. **Artificial intelligence (AI):** Here the bioinformatics functions impersonate the brainpower of the human with computers. Its usefulness is understandable by realising the need and importance in whole genome sequencing, sequence reconstructions, and gene finder. It also generates vital tools for data processing.
4. **Data mining:** Data mining is vital for extrapolation and meaningful significant information can be extracted from huge datasets. The multifaceted arrays of data can be used as input and utilise a variety of arithmetical and numerical techniques to unearth surprising inconsistencies, like grouping and bundling algorithms for any process. An example includes gene annotation in whole genome sequence, domain, and motif discovery. Similarly, mass spectroscopy can classify proteins, although hindrance in data mining may come from variances in complexity, scale, number and the lack of an accepted ontology.
5. **Soft computing:** Many devices are upgraded for storing the biomedical information; still computers have its significant role and hold a special place with researchers and biologists. This has the unique progress of expressing the data of the gene. In addition, it also expresses the bioinformatics data. This evolves with neural network model and artificial neural networks. Similarly, this is the easiest and reliable method for analysing the process. The ultimate factor of this method is genomic and proteomic applications. This is useful for the scientists to do the experiments that result in a vast amount of data.
6. **Simulation and modelling:** The term simulation refers to computation, especially for advanced algorithms and softwares, which are used for curation and analysing the sequence, functions, and structures. Computer simulation is consistent, accommodating, and movable method to support the information. The properties of biomolecules, their interaction like protein–ligand interaction, drug target analysis, enzyme catalysed reactions and protein folding are very well understood with simulation-based methods. Thus, the working of these molecules mimics the actual physiological events. These simulation studies rely on mathematics, physics, biophysics, and chemistry. Like, quantum mechanics and molecular mechanics (QM and MM) and lively mock-ups of proteins are amalgamated with multiple-scale approaches like diffusion models with cellular automata (CA) in brain tumour replication experiments. Another

example of antifungal protein chitosanase/glucanase, soil bacterium *Paenibacillus* sp. shows binding with potential compounds at C-terminus of proteins, which became possible with ligand dockings and free energy estimates (affinity predictions).

Drug development has benefitted a lot from simulation studies. It is well-known that drug discovery and progress require almost 10–15 years, which is not even costly, but also labour intensive. The time and cost input can be significantly reduced with use of computational tools. Drug development is one of the foremost goals of bioinformatics and is popularly known as Computer-Aided Drug Designing (CADD) (Dong & Zheng, 2008; Macalino et al., 2015). It starts with in-silico structure-activity relationships (SAR) studies, which include binding approximation of potential drug molecules on various target sites, which could be: either enzymes, receptors, ion channels, or transporters, inside the cellular membranes. Computers are used in depiction of shapes of reference molecules 2D (2-dimensional) and 3D (3-dimensional) structure, followed by evaluation of their active pharmacophores (important groups involved in binding) and strength of these interactions at target sites, termed 'molecular docking' (Looger et al., 2003).

Different computational softwares used for docking are AutoDock 4, FLIPDock, Vina, SwissDock, UCSF DOCK, FRED, EADock, SWISS MODEL, LOMETS, PatchDock, HADDOCK 2.2, FIND SITE and ClusPro, etc. (Grosdidier et al., 2011; Pagadala et al., 2017). Some of these are paid; however, AutoDock 4 and FRED are freely available. These SAR studies generate large amount of data and can be useful in predicting a lead compound, which has the potential to develop as a drug in future. The success stories include—Sotalol (brand name Betapace) is used to treat a type of fast heartbeat called, sustained ventricular tachycardia. It slows the heartbeat by acting on potassium channels (Brugada et al., 1990). Similarly, Amlodipine (brand name Norvasc) is used for high blood pressure and coronary artery diseases, by inhibiting calcium ion influx across cell membranes (Calcium channel blocker) (Fares et al., 2016). Similarly, Daliresp® (Roflumilast) is a phosphodiesterase (PDE) 4 inhibitor and is used for treating chronic bronchitis, psoriasis, and neuroinflammation by reducing inflammation (Dong & Zheng, 2008).

Although this method has its own limitations, as side effects may not be entirely be predicted and thus requires subsequent validation with actual laboratory experiments and clinical trials.

Image processing: It is essential and assists the biologists and scientists to view their research. This method displays every stage of accomplishments practically.

With the various steps/techniques involved in bioinformatics, we can now proceed with application of bioinformatics.

4 Application of Bioinformatics

Genome Applications: It starts with DNA sequencing, genome assembly, annotation of genes and prediction of gene function (based on the similarity to known genes), sequence analysis for comparative exploration, evolutionary studies, etc. Algorithms such as BLAST, Clustal W and FASTA provide clarification in sequence investigation and examination. This has benefitted the proteomics, transcriptomics, and metabolomics studies. Similarly, functional genomics studies involve RNA-sequence alignment and differential expression analysis. Comparative genomics and computational evolutionary biology shed light on major events in evolution and divergence. Besides them primer designing, restriction enzymes map analysis, RNA fold, dot plot are other genome-based applications.

In predicting protein structure: The RCSB PDB (Research Collaboratory for Structural Bioinformatics Protein Data Bank) provided the first open access digital platform for researchers and is available at 'https://www.rcsb.org/'. It supports retrieval of 3D-structure data of biological molecules, including proteins and involves protein sequence retrieval, followed by virtually establishing the similarities with sequences of known structures, present in the PDB (homology modelling) (Berman et al., 2000).

Biomedical: In the biomedical field, bioinformatics tools have an over powering effect on the understanding of genome, molecular medicine, personalised medicine, and preventive medicine. Novel information on the molecular mechanism of any ailment makes it easier to efficiently treat and prevent the disease. This makes it easier to investigate genes straightforwardly associated with numerous diseases. For all ailments, alike drug is given to patients, but different people have different genotype, so it is important to consider the variations, even subtle but significant (called single nucleotide polymorphism, SNP) in patients' response to drug. It is not exaggeration that if DNA profile of a patient is analysed, the medication would be as unbeaten and efficient as possible, especially in chemotherapy. Like Tamoxifen, (commercially known as Nolvadex), is used in breast cancer treatment, which works by binding on oestrogen receptor in breast tissue. This drug works when a person is positive for oestrogen receptor (ER +ve) and is ineffective in patients lacking this receptor (ER -ve). Similarly, polymorphisms (different forms) of gene for human leukocyte antigen (HLA) alters the immune response and metabolism of drugs. Many other drugs effectiveness has been found related to the genotype and often studied under genome-wide association studies (GWAS).

This proves that success rate in treatment depends on patient's sensitivity/insensitivity to particular drugs, which is sometime dependent on variable forms of genes. This, jeopardy of failure during therapy can be lessened by performing GWAS tests beforehand.

Human microbiome and metagenomics: The word 'microbiome' refers to the entire population of microbes that live inside any specific ecosystem. The human microbiome refers to all the microorganisms living inside human body, without causing the disease, including intestine/gut and includes various bacteria and fungi. Their gene and environmental interactions affect the human physiology. The variations in terms of dissimilar bacterial populations in different humans create a 'unique microflora', which has been studied in population groups. Like autistic children, harbour significantly fewer types of gut bacteria than healthy children do (e.g. *Prevotella* and *Coprococcus* species) (Kang et al., 2013). *P. copri* is involved in breakdown of protein and carbohydrate foods.

Preventive Medicine and Gene Therapy: Preventive medicine focus is on general health, well-being and simultaneously preventing diseases, disability and death (Gochhait, & Omale, 2018). In some cases, the patient's body becomes a laboratory for drugs trial, as gene responsible for disease is still not known, or it could be a multi-gene disorder. Our genetic makeup, environmental conditions, disease agents, lifestyles and genetic predisposition decide the probability of acquiring a disease in future and many people die every year from preventable diseases. These could be chronic respiratory diseases, cardiovascular disease, diabetes, certain infectious diseases, etc. Here, genetic testing can also be done to screen for mutations for disease-related gene polymorphism (single nucleotide polymorphism or SNP) that cause genetic disorders or are predisposed to certain diseases like certain type of cancers or diabetes.

Similarly, gene-editing tools, can be useful in correcting the genotype and alleviating the disease phenotype. Although method is still under research, optimistically it would soon become a reality.

Forensic Analysis and Bioinformatics: Forensic analysis is largely based on DNA-related data, which is also used for personal identification and relatedness. Genomic tests are extensively used as legal evidences in paternity disputes, cadaver recognition, insurance business frauds, and other crimes. This is the reason that many countries are instituting forensic databanks of frequent/serial lawbreakers and criminals. It is possible due to DNA sequencing, evidence orderliness, and machine learning algorithms to establish a Probabilistic Graphical Model (Bayesian networks) (Bianchi & Liò, 2007). The fingerprints, DNA samples, retinal/iris scan (unique patterns of a person's retina blood vessels) and tongue prints are various methods, which are equally effective like signature verification, voice recognition, and face recognition (Radhika et al., 2016). Their biometric databases and similarity searches are possible due to bioinformatics only.

Microbial Genome and Climate change studies: This refers to exploring the details of the genetic material of microbes and isolating the genes, which give them an unparalleled ability to survive in extreme conditions. This application can have ample implications in the improvement of ecosystem, well-being, energy and industrial benefits. A well-suited example is *Pseudomonas putida*, a bacterium with synthesised genome, where DNA of four different species is combined to develop a petroleum-degrading phenotype during notorious oil-spill in ocean in 1980, which successfully cleaned the hydrocarbons floating in water and putting underwater animals at risk. In 1994, the US Department of Energy initiated the Microbial Genome Project, for the sequencing of microbes useful for environmental cleanup, energy production, industrial treatment, and minimising toxic waste. Similarly, projects that are more ambitious can be planned like global climate change, which is chiefly due to increasing levels of carbon dioxide emissions and one way to reduce atmospheric CO_2 is possible by exploring the genomes of microorganisms that uses CO_2 for carbon.

Biotechnology*:* There are many uses of bioinformatics to fasten research in the field of biology, like automated sequencing of genome, gene mapping, protein configuration, organism identification, drug development and vaccine design. Still, there is more scope as biotechnology field is a side street. Few examples are discussed below-

Crop Improvement: It is agreed that the growth in population led to worldwide temperature changes and resulted in declined crop yields. Thus, a major challenge is that people should not die of starvation and this is feasible with a sustainable agricultural production model. Here, comparative genomics aid to understanding genes functions, across plant species. In order to achieve high-quality crops in a short period, bioinformatics databases are used to design new technologies and varieties with better productivity.

Veterinary Science: Mere sufficient food consumption is not adequate for people to survive and stay fit. Our health is dependent on nutrients up take from food. At times, for consuming nutrients, people today depend on livestock as well. A great success achieved in modifying the animal's genotype through bioinformatics, which reduced the risk of possible infection and increasing production.

Innovation in examining the animal species helps in understanding the system genetics of complex traits and provides accurate prediction. Specifically, focuses was given on sequencing genome of animals including—horse, cow, pigs, and sheep. This led to the development in total production, as well as health of livestock. Moreover, bioinformatics has helped researchers in discovering new tools for the discovery of vaccine targets.

Canine cancer approximately affect one in every three dogs and exist as one of the leading causes of death, despite advances in conventional therapies. The cancer pathophysiology is linked with alteration of cellular gene expression, and bioinformatics toolbox promises to put forward innumerable insights into molecular mechanisms, diagnostics and novel therapeutic interventions for cancer. In dogs, osteosarcoma (OSA) affect the lives of 85–90% of affected, within two years of identification, despite uncompromising surgery and chemotherapy. Here, the death is more frequently due to metastasis and thus studying the transcriptome of tumours in bones, scientists have identified a possible therapeutic target. OSA cells express a protein, called her2/neu, on their cell membrane, and metastatic cells express this protein at much higher levels than tumour primary cells, and thus a target for immunotherapy. Scientists created a vaccine, which improves the immune response with obliteration of her2/neu OSA cells. In a first phase of clinical trial, this vaccine show significant improvement over other treatment options.

Another example is equine metabolic syndrome (EMS), an endocrine disorder in horses is linked with obesity, resistance to insulin. Studies show that this syndrome is not only linked with carbohydrate intake, lack of exercise, but there is also a susceptible genotype in horses.

All these examples suggest the wide and ever-growing applications of bioinformatics in biological science and medicine.

5 Conclusions and Future Prospects

Human evolution came a long way, and credit goes to advances in important fields under biotechnology, as diagnostics, drug inventions, clinical health, and agriculture have heightened our financial and social standards. Although a lot is achieved, but still bioinformatics can further help biotechnology in reaching new heights, and helping humankind, but we must set the ethical boundaries and inventions should happen within moral limits. Thus, besides inventions, a string surveillance and regulatory system is need of the hour. Industry practitioners and academicians have to look forward towards the adaptation of AI (Artificial Intelligence) in gaining momentum through big data analysis, machine learning, social media analysis, algorithm decision-making, simulation modelling, and other techniques that is used for bioinformatics visibility in the global market (Varsha et al., 2021).

References

Altschul, S. F., Gish, W., Miller, W., Myers, E. W., & Lipman, D. J. (1990). Basic local alignment search tool. *Journal of Molecular Biology, 215*(3), 403–410. https://doi.org/10.1016/S0022-2836(05)80360-2

Avery, O. T., MacLeod, C. M., & McCarty, M. (1944). Studies on the chemical nature of the substance inducing transformation of pneumococcal types. *The Journal of Experimental Medicine, 79*(2), 137–158.

Berman, H. M., Westbrook, J., Feng, Z., Gilliland, G., Bhat, T. N., Weissig, H., Shindyalov, I. N., & Bourne, P. E. (2000). The Protein Data Bank. *Nucleic Acids Research, 28*(1), 235–242. https://doi.org/10.1093/nar/28.1.235

Bianchi, L., & Liò, P. (2007). Forensic DNA and bioinformatics. *Briefings in Bioinformatics, 8*(2), 117–128. https://doi.org/10.1093/bib/bbm006

Brugada, P., Smeets, J. L., Brugada, J., & Farré, J. (1990). Mechanism of action of sotalol in supraventricular arrhythmias. *Cardiovascular Drugs and Therapy, 4*(Suppl 3), 619–623. https://doi.org/10.1007/BF00357040

Dong, X., & Zheng, W. (2008). A new structure-based QSAR method affords both descriptive and predictive models for phosphodiesterase-4 inhibitors. *Current Chemical Genomics, 2*, 29–39. https://doi.org/10.2174/1875397300802010029

Fares, H., DiNicolantonio, J. J., O'Keefe, J. H., & Lavie, C. J. (2016). Amlodipine in hypertension: A first-line agent with efficacy for improving blood pressure and patient outcomes. *Open Heart, 3*(2), e000473. https://doi.org/10.1136/openhrt-2016-000473

Franklin, R. E., & Gosling, R. G. (1953). Molecular configuration in sodium thymonucleate. *Nature, 171*(4356), 740–741. https://doi.org/10.1038/171740a0

Gochhait, S., & Omale, D. (2018, June). Analytical solution to the Mathematical models of HIV/AIDS with control in a heterogeneous population using homotopy perturbation method (HPM). *Advances in Modelling and Analysis A, 55*(1). ISSN 1258-5769.

Gochhait, S., et al. (2021). Data interpretation and visualization of COVID-19 cases using R programming. *Informatics in Medicine Unlocked, 26*(6). ISSN: 0146-4116.

Grosdidier, A., Zoete, V., & Michielin, O. (2011). SwissDock, a protein-small molecule docking web service based on EADock DSS. *Nucleic Acids Research, 39*(Web Server issue), W270–277. https://doi.org/10.1093/nar/gkr366

Hershey, A. D., & Chase, M. (1952). Independent functions of viral protein and nucleic acid in growth of bacteriophage. *The Journal of General Physiology, 36*(1), 39–56.

Kang, D.-W., Park, J. G., Ilhan, Z. E., Wallstrom, G., LaBaer, J., Adams, J. B., & Krajmalnik- Brown, R. (2013). Reduced incidence of Prevotella and other fermenters in intestinal microflora of autistic children. *PLoS ONE, 8*(7), e68322. https://doi.org/10.1371/journal.pone.0068322

Looger, L. L., Dwyer, M. A., Smith, J. J., & Hellinga, H. W. (2003). Computational design of receptor and sensor proteins with novel functions. *Nature, 423*(6936), 185–190. https://doi.org/10.1038/nature01556

Macalino, S. J. Y., Gosu, V., Hong, S., & Choi, S. (2015). Role of computer-aided drug design in modern drug discovery. *Archives of Pharmacal Research, 38*(9), 1686–1701. https://doi.org/10.1007/s12272-015-0640-5

McPherson, J. D., Marra, M., Hillier, L., Waterston, R. H., Chinwalla, A., Wallis, J., Sekhon, M., Wylie, K., Mardis, E. R., Wilson, R. K., Fulton, R., Kucaba, T. A., Wagner- McPherson, C., Barbazuk, W. B., Gregory, S. G., Humphray, S. J., French, L., Evans, R. S., Bethel, G., … Max-Planck-Institute for Molecular Genetics. (2001). A physical map of the human genome. *Nature, 409*(6822), 934–941. https://doi.org/10.1038/35057157

Pagadala, N. S., Syed, K., & Tuszynski, J. (2017). Software for molecular docking: A review. *Biophysical Reviews, 9*(2), 91–102. https://doi.org/10.1007/s12551-016-0247-1

Porter, T. M., & Hajibabaei, M. (2021). Profile hidden Markov model sequence analysis can help remove putative pseudogenes from DNA barcoding and metabarcoding datasets. *BMC Bioinformatics, 22*, 256. https://doi.org/10.1186/s12859-021-04180-x

Radhika, T., Jeddy, N., & Nithya, S. (2016). Tongue prints: A novel biometric and potential forensic tool. *Journal of Forensic Dental Sciences, 8*(3), 117–119. https://doi.org/10.4103/0975-1475.195119

Sharma, A., Ghosh, D., Divekar, N., Gore, M., Gochhait, S., & Shireshi, S. (2021). Comparing the socio-economic implications of the 1918 Spanish flu and the COVID-19 pandemic in India: A systematic review of literature. *International Social Science Journal, 71*, 23–36. https://doi.org/10.1111/issj.12266

Varsha, P. S., Akter, S., Kumar, A., Gochhait, S., & Patagundi, B. (2021). The impact of artificial intelligence on branding: A bibliometric analysis (1982–2019). *Journal of Global Information Management (JGIM), 29*(4), 221–246. https://doi.org/10.4018/JGIM.20210701.oa10

Venter, J. C., Adams, M. D., Myers, E. W., Li, P. W., Mural, R. J., Sutton, G. G., Smith, H. O., Yandell, M., Evans, C. A., Holt, R. A., Gocayne, J. D., Amanatides, P., Ballew, R. M., Huson, D. H., Wortman, J. R., Zhang, Q., Kodira, C. D., Zheng, X. H., Chen, L., & Zhu, X. (2001). The sequence of the human genome. *Science, 291*(5507), 1304–1351. https://doi.org/10.1126/science.1058040

Watson, J. D., & Crick, F. H. (1953). Molecular structure of nucleic acids; a structure for deoxyribose nucleic acid. *Nature, 171*(4356), 737–738. https://doi.org/10.1038/171737a0

CHAPTER 2

Artificial Intelligence in Biological Sciences: A Brief Overview

Uma Dutta, Nikhil Danny Babu, and Girish S. Setlur

1 Introduction

Artificial intelligence (AI) is the term used to broadly describe intelligence demonstrated by machines. In the natural world, humans and other animals display intelligent behavior in the sense of navigating their environments and solving problems to achieve an end goal like finding a

U. Dutta (✉)
Department of Zoology, Cotton University, Panbazaar, Guwahati, India
e-mail: uma.dutta@cottonuniversity.ac.in

N. D. Babu · G. S. Setlur
Department of Physics, IIT Guwahati, Guwahati, India
e-mail: danny@iitg.ac.in

G. S. Setlur
e-mail: gsetlur@iitg.ac.in

G. S. Setlur
Mehta Family School of Data Science and Artificial Intelligence, IIT Guwahati, Guwahati, India

S. Dutta and S. Gochhait (eds.), *Information Retrieval in Bioinformatics*,
https://doi.org/10.1007/978-981-19-6506-7_2

mate or hunting prey, etc. Even micro-organisms have been observed to demonstrate a certain level of intelligence at the population level in forming complex colonial structures and communicating using chemical signals to overcome adverse situations. In general a system either living or non-living can be said to possess intelligence if it shows complex adaptive behavior targeted toward achieving an end goal. We do not consider our computers that we use for almost everything nowadays to be intelligent because computers simply execute a task given a set of instructions by an intelligent user. But what if we teach computers to make decisions internally without a human user telling it step by step what is to be done. We would be able to solve a host of complex problems that are simply too tedious for the human mind.

Artificial intelligence entered mainstream academia in 1956 when John McCarthy coined this term at a workshop in Dartmouth college. This field was founded on the assumption that human intelligence can be precisely quantified mathematically such that a computer can be made to simulate it. This field arose out of a need to understand how the human brain processes information. In the early days, several approaches to AI were studied, for example, research in neurobiology and cybernetics as well as symbolic AI that considered the idea that intelligence can be reduced to symbol manipulation. However, from the twenty-first century onwards it is evident that machine learning has become the most successful approach to AI research so much so that the term artificial intelligence has become synonymous with machine learning using neural networks.

With the advent of faster computers and increasing access to large amount of data has enabled great advances in machine learning, and data intensive deep learning methods now dominate AI research and applications. AI has come a long way since its inception and now there are machine learning algorithms that classify images with greater than 97% accuracy and there are AI programs that can beat the best human Go player on the planet. AI-based models now influence almost every aspect of our lives from online shopping, entertainment, to diagnosing illnesses. In this chapter, we will discuss how AI/machine learning has influenced modern medicine as well as research in the life sciences. In the following sections, we shall briefly discuss some of the recent developments at the interface of AI and life sciences. The objective of this chapter is mainly to incite interest in the reader about this topic and not to serve as a technical reference material. So keeping this in mind we brush under the carpet the technical details and broadly discuss the state-of-the-art applications of AI-based machine learning models in the health and life sciences.

2 Basics of Machine Learning (ML)

2.1 *Building Blocks of a Neural Network*

Machine learning is the most successful approach to AI. It is basically a computer algorithm that learns through experience using data and is usually optimized for a certain task, for example image recognition or predicting customer behavior, etc. One analogy to this that can be taken from daily life is that of a pair of new boots that feel uncomfortable and cause pain when worn initially but on repeated usage for a couple of days the boots "learn" the shape of the wearers feet and adapt for a comfortable fit. In machine learning, the computer program is trained with sample data called 'training data' to make predictions or decisions without being explicitly programmed to do so. Machine learning is usually implemented using artificial neural networks (NNs). Artificial neural networks were initially developed to mimic the biological brain. A neural network consists of several artificial neurons and connections between these neurons. These artificial neurons are meant to mimic a biological neuron. The construction of the network is such that the neurons are arranged in layers. Every neural network should at least have two layers, an input layer into which the input data is given and an output layer that gives the final output of the NN. Additional layers also called hidden layers can be included in between as needed and this gives a "depth" to the neural network and the popular terminology 'deep learning' is associated with neural networks that have several multiple hidden layers. The hidden layers enable the NN to perform higher levels of abstraction, for example in image recognition it is the hidden layers that detect the edges and features in the image that helps in accurately identifying the image. The number of neurons in a single layer is associated with the 'width' of the neural network. Enhancing the computational efficiency of a neural network mainly involves finding a balance between the width and depth of the network. Each neuron is a mathematical function called the activation function ($f(z)$) that takes an input value and gives an output value. The activation functions used are nonlinear functions. To simply put it, a neural network is a nonlinear function of many variables that depends on many parameters as we shall see below. A linear function has the shape of a straight line. Any function that is not a simple straight line when plotted is termed as a nonlinear function. It is necessary to use nonlinear functions in the neural networks to get the higher levels of abstraction needed for learning to take place.

The strength of each of the connections between the neurons in adjacent layers is quantified by assigning a number called a "weight" to each of the connections. Now consider a neuron in one of the hidden layers. The outputs of the neurons in the preceding layer are multiplied by their assigned weights and all of this is summed up over the number of neurons in that layer and an additional parameter called a "bias" is added to this weighted sum, and this is fed as input to the activation function of the neuron under consideration. This is how information flows from layer to layer in the neural network. This process is called "feedforward" pass through the network. So far we have encountered four parameters that are important for constructing a neural network. We can think of these parameters as knobs that one can adjust to design a neural network of the desired size and configuration. Summarizing these parameters, they are:

Width: The number of neurons in each layer. This can vary from layer to layer.
Depth: The number of hidden layers, i.e., the number of layers other than the input and output layer.
Weights: The strengths of the connections between the neurons.
Bias: This is a parameter associated with each neuron. It is basically a number that is added to the input of the neuron. It is also referred to as an offset.

This encapsulates the basic structure of a neural network. Now we come to the interesting part, how does the learning happen? How do we go from here to accurately diagnosing Alzheimer's from fMRI scan images? How do we get it to make accurate predictions? This requires optimization of the neural network and there is no given prescription for this, it depends on the particular task at hand.

3 How Do Neural Networks Learn?

The human brain is truly incredible. A child who has seen a lion only in pictures of a drawing book can immediately identify a real lion on a trip to the zoo. Even such a simple task is very challenging for a computer program. It requires a large dataset to train the neural network to produce the correct output. Neural networks learn by training. The procedure of training a neural network is what is known as machine learning. It typically requires a huge amount of data to train a neural network and obtaining the relevant data falls in the domain of data mining and data science. There are three main techniques used in machine learning. These are **supervised learning**, **unsupervised learning**, and **reinforcement learning**.

Supervised learning: In this method, the model is trained with supervision analogous to a student learning in the presence of a teacher. The input data is processed and labeled. Suppose we have a large collection of images of animals. The task for the neural network is to correctly identify the animal in each image. The input data is divided into a training set and a test set. The data in the training set as well as the correct output (names of the animals) will be supplied to the model and it will be trained to correctly identify the animals in the training set. Once the training is complete, the model will be tested by supplying images from the test set and it will be able to correctly identify the images it has never seen before. This technique is widely used for problems of classification and regression.

Unsupervised learning: In this method, the model infers patterns from unlabeled input data. This method does not require supervision and the algorithm discovers patterns in the data by itself. Suppose we have a large collection of images of animals and we want to group the images into different categories, like reptiles, birds, etc. This task can be accomplished using unsupervised learning. This method enables the neural network to identify hidden patterns in the data. This can be useful when dealing with an unfamiliar dataset for which the labels, i.e., the corresponding output data are not available. Unsupervised learning is used for clustering and association problems.

Reinforcement learning: This is a feedback-based method of machine learning in which the program learns from the outcome of its actions. It receives positive feedback for correct actions and negative feedback for incorrect actions. This method does not require any labeled data and is used to design AI that needs to interact with its environment and navigate through it like in the case of robots, game playing AI, self-driving cars, etc.

We will briefly discuss only supervised learning as it is the one most commonly used in several applications of AI in the biological sciences.

The other two methods are beyond the scope of this chapter. In supervised learning, the labeled dataset is split into a training dataset and a test dataset which is used to benchmark the trained model. During the course of the training, the architecture of the neural network remains the same, i.e., the depth and width parameters of the network are not adjusted, but the weights and the biases (initially assigned randomly) will keep changing until their optimum values are obtained. For this reason, the weights and biases are called hyperparameters. In simple words, what the neural network learns is the optimal values of all the weight and biases that allow it to make accurate predictions. The concept of a cost function is needed to understand this process. It is basically a mathematical function of the set of weights (w) and biases (b) denoted by $C(w, b)$ and it is proportional to the square of the difference between the correct output/labels of the training data $F(y^{\text{train}})$ and the output produced by the neural network $F_{w,b}(y^{train})$. That is, $C(w, b) \propto F_{w,b}(y^{\text{train}}) - F(y^{\text{train}})^2$. The idea is to find the set of weights and biases for which the cost function $C(w, b)$ is minimum. The cost function is like a penalty for the NN and the target is to achieve the minimum possible penalty. This is accomplished using "gradient descent" algorithm and backpropagation. Gradient is the derivative of the cost function with respect to the weight dC/dw. If the cost increases with increasing weight the gradient will be positive, on the other hand, if the cost decreases with increasing weight the gradient will be negative. The model needs to know whether to increase or decrease the weights in order to minimize the cost function, the negative of the gradient ($-dC/dw$) shows exactly this. Now the model knows in which direction to move the weights but we must specify by what amount it must change the weights and this is decided by the learning rate parameter η. The weights are updated according to the following formula $w = w - \eta|dC/dw|$ till the minimum of the cost function (where $dC/dw = 0$) is reached. We have just briefly covered the very basic introduction to gradient descent but in fact in practice several gradient descent techniques are employed like 'stochastic gradient descent,' 'batch gradient descent,' 'mini batch gradient descent,' etc. In practice the situation is more complex than what we have described as the cost function can have multiple minima and the gradient descent procedure can get stuck around a local minima but the model should reach the global minimum of the cost function in order for it to be successfully optimized. This issue can arise if the initially assigned weights are near a local minimum of the cost function. AI engineers use "backpropagation" to ensure that the global minimum of the cost function is reached by the gradient descent algorithm. The cost function is calculated at the output layer and this information should be backpropagated to the previous layers and all the weights associated with the neuron connections all the way up to the input layer should be adjusted using the gradient descent formula. This reverse flow of information in the neural network is termed backpropagation. The mathematical details of this procedure are beyond the scope of this chapter, and the interested reader can refer to any of the excellent introductory books and articles on machine learning (Alpaydin, 2020; Baştanlar & Ozuysal, 2014; Kubat, 2017).

4 Applications of AI in the Life Sciences

With the rapid progress in data analytics techniques and computing power artificial intelligence has become more accessible and it has made its impact in a wide array of disciplines. In recent years, AI has made a profound impact in health care, life sciences, and bioinformatics. Popular AI tools such as deep learning neural networks are used in cancer, neurology, and cardiology research. AI is no longer restricted to the domain of computer science, and it has become paramount for researchers and practitioners in other areas of science and arts to get exposed to this field. In the following sections, we shall give a brief overview of some of the recent applications of AI in biology and healthcare.

4.1 Using Deep Neural Networks to Understand the Biological Brain

The biological brain is organized such that it has specialized areas for performing different tasks. For example, there is an area in the brain that recognizes objects but there is another region specialized for recognizing faces in particular. Neuroscientists have struggled to understand why the brain has specialized regions for various tasks. Now deep learning neural networks are providing insights into why such specialization is so effective. More and more neuroscientists are beginning to use AI to shed light on the inner workings of the human brain. In image recognition traditional neural networks as described in the previous section run into difficulties when a shifted version of the input image is encountered. It does not perform well if a picture of a cat curled into a ball is given to it and an image of a cat fully stretched is given to it. It tends to overfit the data and hence underperforms when images of the same object in different positions is given to it. So the NNs are not translational invariant and are not efficient in feature recognition. Also for high-resolution images traditional NNs would require a large number of neurons and would require fitting of an impractical number of weights and biases. A good AI model should be able to identify a cat wherever it occurs in a picture and whatever position the cat may be in, like running, sitting, or jumping. Convolutional neural networks (CNNs) overcome all these problems.

In a convolutional neural network, filters are applied to the input image in the hidden layers. For example, a filter that detects eyes could be applied to the image and it detects how many times and in what locations an eye is present and creates a feature map. Similarly, filters that detect other features are applied and feature maps are created. These feature maps are then supplied as input to an activation function (usually reLU). The output of the activation function determines whether a certain feature is present at a location in the image or not. Adding more layers of filters and feature maps will result in a deeper CNN that is capable of detecting more abstract features. Pooling layers are used in between convolutional layers, in which the largest values in the feature maps are selected and passed onto the subsequent layers. This is called max pooling and improves efficiency of feature detection. Finally, a fully connected layer (all the neurons in the layer are connected to the previous layer) performs the image classification in the end. A detailed explanation of the CNN architecture can be found in (Albelwi & Mahmood, 2017). In recent works by Bonnen et al. (2021) and Zhuang et al. (2021), the authors have used a deep convolutional neural network (CNN) to classify images. They observed that the basic shape and outline of the objects in the images were captured by the earlier stages of the neural network while the more complex features were captured at the deeper stages in the network. This is very similar to how the primate visual system works. In the primate visual system, the pathway responsible for recognizing people, places, and other objects is the ventral visual stream. The signals from the eyes travel sequentially through the lateral geniculate nucleus to the V1, V2, and V4 centers of the primary visual cortex and reach the inferior temporal cortex. In the initial stages of the pathway, the basic features like edges and shapes are detected and more complex features like eyes and nose, for example, are detected in the later stages of the ventral visual stream. The specificity in the functional match between the deep convolutional neural net and the primate brain has gotten computational neuroscientists excited about the prospect of using AI to understand human brain function.

We do not have a good understanding of how the brain processes auditory information. How the human brain processes sound in the auditory cortex is still very much a open problem in neuroscience. In a recent groundbreaking work by Kell et al. (2018) the authors designed a deep neural network and optimized it to distinguish between two types of sound: speech and music. Their goal was to find the best deep convolutional neural net architecture that could perform this task efficiently with minimum resources. The input given to the model was an audio clip of speech with background noise or music with background noise. The task given to the network was to identify the word spoken at a particular time in the first case or to identify the genre of music playing in the latter case. The researchers experimented with three possible architectures: (*a*) Two separate networks specialized for each task with no shared layers. (*b*) A network that is branched in the middle. Few of the initial layers are shared. In the deeper part, the network is separated and higher-order processing for each task occurs separately. (*c*) All the layers are shared and process both the tasks, only the output layer is separate. The best performing architecture was the one which was branched in the middle. It was able to detect speech and recognize music efficiently using minimum computing resources.

The separate networks with dedicated pathways for speech and music tasks are the most accurate but it is also computationally expensive. The network with separate speech and music pathways and a shared front end was shown to be the optimum architecture, and it also agreed with predictions that the deeper layers of auditory cortex have distinct regions for processing speech and music. Moreover, this network made human-like error patterns and predicted auditory cortical responses. The researchers were able to infer the hierarchical organization in the human auditory cortex from predictions made by a neural network. A deep neural network was also successfully used to model the olfactory system in fruit flies which is one of the most well-studied systems in neuroscience (Wang et al., 2021). Artificial intelligence continues to make huge strides in helping us understand better the inner workings of our own minds.

4.2 *AI in Medical Diagnosis*

The incredible success of AI in image recognition means that it can be put to use with confidence for important applications like medical diagnosis. Early detection of diseases like cancer or Alzheimer's disease could make a huge difference in the outcome of the patient's treatment. If AI can be trained to predict the onslaught of cancer or early stages of Alzheimer's by looking at the CT scan or fMRI images before it becomes apparent to a trained human doctor's eyes it could go a long way in saving a patient's life. Several research groups all over the world are already tackling this problem and huge strides have been made in the last five years. In the not so far future AI medical diagnostics will be used to save thousands of lives every year. But there are several challenges that need to be overcome first. Convolutional neural networks (CNNs) are extremely adept at learning patterns from data hence they are used extensively in chess playing AI, computer vision, and even in language processing AI. But they require a considerable amount of labeled training data and training time. There are millions of labeled images of common objects and animals on the internet but there is a shortage of correctly labeled and medically accurate training data for diagnostic purposes. Another major limiting factor is that CNNs are efficient only when the input data is two dimensional or planar. The algorithm will run into errors if the objects are in 3D, for example a CT scan image of body tissue.

A team of researchers have developed a framework for building neural networks that can detect patterns on any kind of irregular surface. These are called equivariant convolutional neural networks. These equivariant CNNs can be used to learn the patterns in inhomogeneous curved surfaces of body tissue like the heart, brain, lungs, and other organs. Equivariance is similar to the idea of covariance in physics which means that the physical laws do not change with change of reference frame. That is, the laws of physics remain same for a stationary observer as well as for a moving observer. Although the values of measurements made in different frames can change, they can be transformed into each other while preserving the relationship between different quantities. Equivariant CNNs are able to learn same patterns or features that occur in different orientations or locations in the image even if the training examples don't contain these different orientations thereby drastically reducing the amount of training data required. This is highly useful for applications like cancer detection as there is a shortage of labeled CT scan data in such cases. Researchers have already obtained positive results in recognizing lung tumors from CT scans using just a fraction of the training data used in traditional NNs (Winkels & Cohen, 2019). The filter used in CNNs is translational invariant in the sense that it learns a pattern in any location of the image in 2D. In irregular curved surfaces the orientation (or gauge) of the filter can change depending on the path it takes to traverse the surface. This will affect the ability of the network to learn the features. By fixing a certain orientation (gauge) of the filter the researchers found a consistent way to transform every other possible orientation into it. Any arbitrary gauge can be chosen initially but the transformation of other gauges into the initial one should preserve the underlying feature. These gauge equivariant CNNs were demonstrated to work on any arbitrary surface (Cohen et al., 2019).

State-of-the-art 3D image processing CNNs were recently used to learn features to diagnose Alzheimer's disease which is a common neurodegenerative disease (Huang et al., 2019). Alzheimer's disease mostly occurs in aged people and can cause severe cognitive impairment and behavioral issues. The common biomarker indicating the onset of Alzheimer's is shrunken hippocampi. Usually by the time the diagnosis is made using MRI scans, the patients already start showing symptoms. Deep learning AI models can be trained using MRI images of early stage Alzheimer's to predict the onset of the disease before it is too late to provide proper treatment to the patient.

4.3 *Decoding Protein Structure Using AI*

Proteins are large complex molecules that are fundamental to many biological processes. Decoding the structure of proteins and how they interact with each other is a challenging problem that has intrigued computational biologists for several decades. The popular 'protein folding problem' involves understanding the relationship between the protein's amino acid constituents and its final 3D structure. In the early days, supercomputers like IBM's Blue Gene were put to this task and still it was a cumbersome and time consuming effort. But in just a couple of decades the rapid progress in AI research has culminated in programs like Google DeepMind's AI called AlphaFold that has made enormous progress in determining a protein's 3D shape from its amino acid sequence. This has transformed bioinformatics and structural biology research. AlphaFold's predictions of some protein structures were identical to those obtained using experimental techniques like X-ray crystallography and cryo-electron microscopy (cryo-EM). Most of the protein structures that we already know were obtained using X-ray crystallography and now cryo-EM is also being used. But this is a tedious process and it takes time to arrive at the 3D protein structure. The use of AI like AlphaFold can significantly speed up this process and can help in accelerating drug design for diseases that results in faster production of vaccines and drugs for new diseases. There are also other models like MaSIF (molecular surface interaction fingerprinting) (Sverrisson et al., 2021) developed by independent research groups that are less complex than AlphaFold but highly successful in predicting protein structure. MaSIF targets the curved and irregular 2D surface of the protein using geometric deep learning using gauge equivariant CNNs which we have seen in the previous section. By examining the surface of the protein, MaSIF can predict binding sites on the surface. The researchers used surface features like surface curvature, electric charge, and hydrophilicity to train the model. The model then learnt to use these features to detect higher-order patterns. This AI significantly speeds up the process if identifying potential fits among a large set of proteins. MaSIF is not capable of predicting induced fit which is when the molecular surface changes shape and chemistry in the vicinity of another molecule to which it can potentially bind to. Researchers are working on designing AI that can predict induced fit and other surface dynamics as well. AI played an important role in our fight against the COVID-19 pandemic that shook the world. In 2020 AlphaFold was used to predict the structure of some SARS-CoV-2 proteins that were not solved experimentally and its prediction of Orf3a protein in the virus ended up matching perfectly with the cryo-EM structure. This played an important role in synthesizing antiviral proteins to fight the virus. MaSIF was used to predict the surface fingerprint of SARS-CoV-2 spike protein.

4.4 AI in Ecology and Conservation Biology

Every year a huge amount of ecological data is being collected by researchers across the world. This includes satellite images, aerial images, audio clips, camera traps, survey data, etc. This data can be used to predict wildlife population movement, invasive species outbreak, changes in vegetation and is even used in the fight against illegal wildlife trafficking. Manually processing and analyzing these huge troves of data is cumbersome and time consuming. Increasingly AI is being used for this purpose and it can detect infrequent patterns and complex structures in the data that is imperceptible to humans. This speedup achieved by AI is necessary as researchers need immediate answers to pressing questions like whether conservation efforts are working. Computer vision algorithms and audio processing algorithms both making use of CNN architecture are commonly used in ecology and conservation. AI is more accessible than ever and inexpensive ready to use AI applications are being developed that can be used by people with no programming knowledge.

5 Limitations of AI

The field of artificial intelligence has come a long way since its inception in the 1950s and now has become an indispensable part of technology. But there is a caveat since we don't yet understand completely the inner workings of a deep learning model. There are several pitfalls and limitations to AI that can easily be exploited to make it give completely wrong predictions even for simple tasks. It seems that artificial intelligence is not that intelligent after all and it should be used with caution with human supervision. Blindly following the predictions of AI can be misleading in some cases.

5.1 *Overfitting*

If a machine learning model gives correct predictions on training data with good confidence but performs sub-optimally on new data used for validation then it means that the model is overfitting the data. This happens when the model starts learning the noisy features in the training data in addition to the useful features. When there are more hyperparameters (weights and bias) to adjust than is necessary the model tends to overfit the data and loses its predictive flexibility when supplied with new data. To avoid overfitting the training process should be stopped just before the cost function starts increasing after an initial decrease, if the training process is continued for longer than necessary, then the model becomes highly specialized to the training data won't be generalized to unknown data. Also selective dropping of some connections (weights) during backpropagation will reduce the number of adjustable parameters and decreases the complexity of the model to avoid overfitting.

Underfitting occurs when the model is not optimized properly and performs poorly with the training data. To avoid underfitting more training examples should be included and the complexity of the model also should be increased. A simple pictorial representation of underfitting, overfitting, and balanced fitting of training data by a machine learning model.

5.2 *Adversarial Examples*

Artificial intelligence can be tricked more easily than one could imagine. A prime example of this is an adversarial example. An adversarial example is input data that is modified to include some subtle noise that causes the machine learning model to completely wrong predictions. This is best illustrated in image recognition models where the corrupted input image looks the same as the correct one to the human eye but the AI recognizes it as a completely different object. This not only affects supervised learning models but can affect reinforcement learning models too. This can be used for malicious purposes as AI security systems can be thwarted using adversarial examples and also self-driving cars can be sabotaged by simply making minor changes to a stop sign such that the AI doesn't recognize it. A basic method to create adversarial examples is the 'fast gradient sign' (Goodfellow et al., 2014) method in which the perturbation or noise added to the input image (data) is $N = \epsilon\ sign(\nabla_x C(w, x, y))$. Here ∇ operator means taking the derivative of a function with respect to any of its parameters and the cost function $C(w, x, y)$ depends on the weights and biases denoted by w, x is the input to the model (pixel values in case of images) and y is the target outputs. The gradient w.r.t x tells us how the cost function changes with change in say the input pixel values. What is needed is just the sign of the gradient as this tells us whether the input pixel values should be increased or decreased. The value ϵ is taken to be very small so that the perturbation or noise added is undetectable by the human users who input the data. The adversarial input is the original input plus the added noise N.

Researchers have developed a few methods to defend against adversarial attacks, one such method is 'adversarial training.' In this method, adversarial examples are included in the training data with their corresponding correct labels as targets and the numbers of training runs are increased. But this takes up more resources and training time and it is not foolproof against more sophisticated attacks. Another method is 'Ensemble Adversarial Training' which uses adversarial examples generated from several other pre-trained models to train the current model of interest (Tramèr et al., 2020). This drastically reduces error rate due to more sophisticated adversarial attacks.

5.3 *Data Bias*

Big tech companies pour in millions of dollars to develop more and more sophisticated deep learning AI models but in the end its performance heavily depends on the quality of data that is used to train the model. Processing the data before feeding it to AI models is turning out be a complex task. Skewed labeling of training data can render the model inefficient in real-world situations even if it performs well in the training process. This is particularly important when dealing with AI used in medical diagnostics as biased or improperly labeled data can cause misdiagnosis and could delay treatment procedures. One of the main difficulties is in weeding out biases in the data that inadvertently creep in due to human error. In a recent incident insufficient representation of black Africans in training of Facebook's image recognition, AI caused it to group Africans and primates into the same group in an embarrassing incident for the social network giant. This dependency on quality and sometimes quantity of data is a limiting factor for the wide applicability of AI.

6 Conclusions

In this chapter, we have given a brief overview of the impact of artificial intelligence (AI) in the biological sciences and bioinformatics (Varsha et al., 2021). Using simple examples and basic terminology, we briefly described the building blocks of AI and the steps that go into implementing a successful model. Without burdening the reader with mathematical detail we discussed what machine learning is and how the learning process in a neural network works. We described how AI is being used to push the frontiers in understanding the working of our brain and also how it has become an indispensable tool in modern medical diagnosis. Some of the most difficult scientific problems of the twentieth century like the protein folding problem have become more tractable with cutting-edge developments in AI in recent years. Exponential progress is being made in this field day by day and it is becoming evident that artificial intelligence together with human curiosity and innovation would be able to tackle the biggest challenges that humankind is faced with.

References

Albelwi, S., & Mahmood, A. (2017). A framework for designing the architectures of deep convolutional neural networks. *Entropy, 19*(6). https://doi.org/10.3390/e19060242. https://www.mdpi.com/1099-4300/19/6/242

Alpaydin, E. (2020). *Introduction to machine learning*. MIT press

Baştanlar, Y., & Ozuysal, M. (2014). Introduction to machine learning. In *MiRNomics: MicroRNA biology and computational analysis* (pp 105–128).

Bonnen, T., Yamins, D. L., & Wagner, A. D. (2021). When the ventral visual stream is not enough: A deep learning account of medial temporal lobe involvement in perception. *Neuron, 109*(17), 2755–2766.e6. https://doi.org/10.1016/j.neuron.2021.06.018. https://www.sciencedirect.com/science/article/pii/S0896627321004591

Cohen, T., Weiler, M., Kicanaoglu, B., & Welling, M. (2019). Gauge equivariant convolutional networks and the icosahedral CNN. In K. Chaudhuri, R. Salakhutdinov (Eds.), *Proceedings of the 36th International Conference on Machine Learning, PMLR, Proceedings of Machine Learning Research* (Vol. 97, pp. 1321–1330). https://proceedings.mlr.press/v97/cohen19d.html

Goodfellow, I. J., Shlens, J., & Szegedy, C. (2014). Explaining and harnessing adversarial examples. *arXiv preprint arXiv:1412.6572.*

Huang, Y., Xu, J., Zhou, Y., Tong, T., Zhuang, X., & ADNI. (2019). Diagnosis of Alzheimer's disease via multi-modality 3d convolutional neural network. *Frontiers in Neuroscience, 13*, 509. https://doi.org/10.3389/fnins.2019.00509. https://www.frontiersin.org/article/10.3389/fnins.2019.00509

Kell, A. J., Yamins, D. L., Shook, E. N., Norman-Haignere, S. V., & McDermott, J. H. (2018) A task-optimized neural network replicates human auditory behavior, predicts brain responses, and reveals a cortical processing hierarchy. *Neuron, 98*(3), 630–644.e16. https://doi.org/10.1016/j.neuron.2018.03.044. https://www.sciencedirect.com/science/article/pii/S0896627318302502

Kubat, M. (2017). *An introduction to machine learning*. Springer.

Sverrisson, F., Feydy, J., Correia, B. E., & Bronstein, M. M. (2021). Fast end-to-end learning on protein surfaces. In *Proceedings of the IEEE/CVF Conference on Computer Vision and Pattern Recognition (CVPR)* (pp. 15272–15281).

Tramèr F., Kurakin, A., Papernot, N., Goodfellow, I., Boneh, D., & McDaniel, P. (2020). Ensemble adversarial training: Attacks and defenses. *arXiv preprint arXiv:1705.07204.*

Varsha, P. S., Akter, S., Kumar, A., Gochhait, S., & Patagundi, B. (2021). The impact of artificial intelligence on branding: A bibliometric analysis (1982–2019). *Journal of Global Information Management (JGIM), 29*(4), 221–246. https://doi.org/10.4018/JGIM.20210701.oa10

Wang, P. Y., Sun, Y., Axel, R., Abbott, L., & Yang, G. R. (2021). Evolving the olfactory system with machine learning. *bioRxiv*. https://doi.org/10.1101/2021.04.15.439917. https://www.biorxiv.org/content/early/2021/04/16/2021.04.15.439917, https://www.biorxiv.org/content/early/2021/04/16/2021.04.15.439917.full.pdf

Winkels, M., & Cohen, T. S. (2019). Pulmonary nodule detection in CT scans with equivariant CNNs. *Medical Image Analysis, 55*, 15–26. https://doi.org/10.1016/j.media.2019.03.010. https://www.sciencedirect.com/science/article/pii/S136184151830608X

Zhuang, C., Yan, S., Nayebi, A., Schrimpf, M., Frank, M. C., DiCarlo, J. J., & Yamins, D. L. K. (2021). Unsupervised neural network models of the ventral visual stream. *Proceedings of the National Academy of Sciences, 118*(3). https://doi.org/10.1073/pnas.2014196118. https://www.pnas.org/content/118/3/e2014196118. https://www.pnas.org/content/118/3/e2014196118.full.pdf

CHAPTER 3

A Review of Recent Advances in Translational Bioinformatics and Systems Biomedicine

Chittaranjan Baruah, Bhabesh Deka, and Saurov Mahanta

1 Introduction

Translational research utilises scientific findings produced in the lab, clinic, or field and turns them into novel therapies and medical care methods that directly enhance human health. Translational research aims to transfer fundamental scientific discoveries into application more

C. Baruah
Bioinformatics Laboratory, Postgraduate Department of Zoology, Darrang College, Tezpur, India

B. Deka (✉)
North Bengal Regional R and D Centre, Tea Research Association, Jalpaiguri, India
e-mail: bhabesh.deka@gmail.com

S. Mahanta
National Institute of Electronics and Information Technology (NIELIT), Guwahati, India

S. Dutta and S. Gochhait (eds.), *Information Retrieval in Bioinformatics*,
https://doi.org/10.1007/978-981-19-6506-7_3

rapidly and efficiently. When it comes to productivity and translating research into new healthcare advancements, it provides a wide range of specialised resources. Research that fosters and supports multidisciplinary collaboration between laboratory and clinical researchers considers the requirements of communities, and discovers and promotes the adoption of best medical and healthcare practices are all part of translational research (Gochhait et al., 2021).

According to the stage in the translation process (from the commencement of research to social application and effect), translational research is categorised. The T-Spectrum (Translational Spectrum) below depicts the many stages of translational research and development. Transcription, translation, and mRNA and protein turnover are all part of gene expression. What happens when this dependence breaks down? (Buccitelli & Selbach, 2020). A post-translational modification (PTM) affects the fate of proteins in eukaryotic cells. Many web-tool predictors for different PTMs are launched to help diagnose and prevent illnesses (Mohabatkar et al., 2017).

Information technology and databases are used in bioinformatics research to solve biological questions. Genomic and proteomic bioinformatics applications are of global importance. The study of genomes is known as genomics or genome research. A whole genome is a collection of DNA sequences that contain the genetic information that has been passed down from one generation to the next over the ages. It is a physical and functional unit of heredity that is passed from parents to offspring via genetic inheritance. To summarise: Genomic research includes the sequencing and analysis of all genetic material in an organism—from genes to transcripts. To put it another way: Proteomics studies all proteins together called the proteome. Beyond genomics and proteomics, bioinformatics is utilised in a wide range of biological disciplines (i.e. metabolomics, transcriptomics). Bioinformatics is a branch of study that aims to understand complex biological processes via the use of computers. Protein structural alterations, activities, and functions are regulated by post-translational modifications (PTMs) in almost every biological process and activity. Understanding cellular and molecular processes begin with protein PTM identification. However, unlike tedious trials, PTM prediction utilising different bioinformatics methods may offer accurate, convenient, and efficient techniques while also providing important information for future studies (Liu et al., 2015). The introduction of next-generation sequencing (NGS) technology has accelerated the identification of prostate cancer biomarkers Prostate cancer diagnosis and prognosis remains difficult despite the deluge of sequencing data. Chen et al. discussed high-throughput sequencing's recent advances in prostate cancer biomarkers (Chen et al., 2013a).

Many scientists now refer to systems biology as the next wave in bioinformatics. Integrating genetic, proteomic, and bioinformatics data creates a holistic perspective of a biological entity. Systems biology may study how a signalling pathway operates in a cell. Systems biology can simulate the genes involved in the process, their interactions, and how changes affect downstream consequences. Bioinformatics may be used in any system that can express information digitally. From single cells to entire ecosystems, bioinformatics may be used. The full "parts lists" of a genome help scientists better understand complicated biological processes. Identifying how these components interact in a genome or proteome is the subsequent stage of intricacy in the research process.

2 Translational Bioinformatics

Translational bioinformatics (TBI) is a new area that applies biological research to patient care and medication discovery. Develop and analyse clinical and biological data to study illness heterogeneity using computer methods. The search for disease gene(s) requires a thorough understanding of the complex network of biological mechanisms involved in disease progression. This chapter aims to outline the biological and clinical data integration strategy. It also explains the key datasets and techniques used in translational bioinformatics to treat illnesses.

Translational bioinformatics focuses on utilising current research to connect biological data with clinical informatics. Translational bioinformatics now covers the biological and healthcare industries, bridging the gaps between ~~the~~ bioinformatics and medical informatics. Translational bioinformatics has made several databases available to researchers. These databases are useful for physicians, biologists, clinical researchers, bioinformaticians, and health care researchers. These databases help biologists comprehend illness management and medication development techniques, which help them, generate novel hypotheses. Gene variations, enzymes, and descriptive genomics databases are examples of translational bioinformatics databases.

2.1 *Categorising Translational Bioinformatics Research*

Translational bioinformatics research can be basically sorted into four categories (Denny, 2014):

- The utilisation of Clinical "big data" or data from electronic health records (HER) for **genomic discovery**
- Regular clinical use of genomics and **pharmacogenomics**
- **Drug discovery** based on OMICS data and development
- Individual-level genetic testing to **address ethical, legal, and societal challenges related to such services.**

Translational bioinformatics integrates biostatistics, molecular bioinformatics, clinical informatics, and statistical genetics (Chen et al., 2013b). The field is quickly developing, and many related topics have been suggested. Among them, pharmacogenomics is a branch of genomics concerned with genetic differences in drug response. This branch is vital for future precision medicine design. Translational bioinformatics is a recent subject that has gained significance in the era of personalised and precision medicine. Based on curating large amounts of scientific literature, TBI may identify erroneous research, derive fresh insights into underlying genetic mechanisms of disease, enhance estimations of pathogenicity of human genetic variants, and find possible new treatment targets. However, interpreting the data to establish a clinical diagnosis or treatment plan is far more difficult than sequencing an exome. Many of the thousands of discovered variations will need to be examined clinically. Some Mendelian disorders require just one variation to be found and examined, such as basic Mendelian disorders. Multivariate analysis will be required for more complicated disorders (such as cancer, diabetes, and neurodegenerative diseases). Getting accurate results requires asking the proper questions about patients and diseases, as well as using the right computational tools. "Translational genomics" is the use of new findings from the Human Genome Project to enhance diagnostics, prognostics, and treatments for complex illnesses.

Biology and information technology have merged to create a new area of translational study called translational bioinformatics (TBI) (Ritchie et al., 2020). Aside from basic DNA sequence alterations, epigenomic data now contains information on methylation and histone modifications as well as DNA methylation (data above the genome). Through information technology, it is possible to acquire and analyse the proteome (the total amount of proteins in the cell, tissue or organism), transcriptome (the total amount of mRNA in a cell), and metabolome (complete set of small molecules, called metabolites, in the cell). Bioinformatics aims to characterise and quantify molecular groups that contribute to an organism's structure, function, or dynamics. This means that every person's OMICS profile should be linked with their clinical observations, medical images, and physiological signals.

3 Bioinformatics Interventions in Translational Research

The National Institutes of Health (NIH) defines translational research as having two areas of translation. There are two ways to achieve this: one is to adapt findings acquired during laboratory and preclinical research to the creation of clinical trials and human studies. It's also important to note that the second area of translation involves research targeted at boosting the adoption of best practices in the community. In translational research, the cost-effectiveness of preventive and treatment methods is equally essential. From the scientist to the user, translational research shifts the emphasis. User involvement is increasingly important in the translational research paradigm. They have an impact on the priorities of academics.

3.1 Translational Biomedicine

Translational biomedical research has recently been a hot topic in the biomedical research field. Translational research seeks to "translate" current biological knowledge into methods and instruments for treating human illness. Beyond that, it's just medical applied research, with its own obscure name. Translational research covers medical genetics, cancer, and cardiology. Currently studied hereditary diseases include microorganisms, plants, animals, and humans. Mendelism established genetics aims to improve our knowledge of basic, preclinical, clinical, epidemiological, and healthcare research understandings are improved.

3.2 Translational Clinical Research

Several recent studies have highlighted a 20-year gap between top clinical research knowledge and its use in our health system. Translational science aims to accelerate the application of scientific discoveries in clinical settings and foster collaboration between researchers and clinicians. Preclinical research, clinical trials, and health technology evaluation, including Alzheimer's disease and dementia prevention and treatment, may all benefit from the findings.

3.3 Translational Stroke Research (TSR)

TSR includes fundamental, translational, and clinical research. Modern methods for evaluation, prevention, treatment, and repair after stroke and various types of neurotrauma are being developed. Basic and clinical scientists and doctors alike may benefit from translational stroke research. This includes neuroscientists and physicians alike. Preclinical and clinical neuroprotective effectiveness differences have become more concerning in translational stroke research.

3.4 Translational Neuromedicine

To deliver new treatments with quantifiable results to patients with neurological disorders, translational neurology studies all technological advancements. Conceived to help those at risk of or suffering from neurological illness convert the vast amount of fundamental neuroscience, neuropathogenesis, and neuroengineering knowledge into treatments and quantifiable benefits, bringing together fundamental and clinical neuroscientists, Translational Neuroscience aims to improve our understanding of brain anatomy, function, and illness.

3.5 Translational Oncology

Work in both the laboratory and the clinic to improve oncology patient care. Clinical trials evaluating new treatment paradigms for cancer are the outcome of Translational Oncology research. In addition, it includes the most sophisticated clinical tests of both traditional and novel cancer treatments. Research in Translational Cancer Treatment promotes facilities for cancer treatment and notable programmes in relevant areas to combined interdisciplinary and translational cancer control companies. Laboratory discoveries are translated into new cancer treatments for patients via translational cancer research (TCR). Because this study often leads to effective treatments for patients as quickly as possible, it is a boon to society. Researchers develop instruments for clinical trials with the use of clinical observations made by doctors. Conversely, doctors utilise clinical observations to guide their efforts.

3.6 *Translational Imaging*

Clinical and scientific applications of biomedical imaging. These methods are being used in clinical investigations to uncover chemical imbalances linked with serious mental disorders and drug addictions. To evaluate traits in animal models, imaging can detect them, and vice versa. These studies attempt to "translate" ideas from micro-imaging laboratories to preclinical settings.

3.7 *Discovery Biology*

In the areas of cancer and neglected diseases, discovery biology performs basic and applied drug discovery research. The field of discovery biology is concerned with basic and practical drug discovery research, especially in cancer and other neglected diseases. Discovery biology services include biosafety and ad hoc virus testing (Yin et al., 2021).

3.8 *Medical Biotechnology*

Medical biotechnology focuses on developing technologies for the health and pharmaceutical industries. Biotechnology in medicine enables researchers and doctors to find new drugs and prevention of diseases. Most medical biotechnologists work in academia or industry. Biotechnologists have discovered novel medicines and developed and tested diagnostic technologies to treat and prevent disease. Depending on their expertise, medical biotechnologists work in academia or industry. Academic biotechnologists help medical researchers conduct tests, whereas industrial biotechnologists create pharmaceutical drugs and vaccines. Medical biotechnology has created microbial insecticides, insect-resistant crops, and environmental cleaning techniques.

3.9 *Orthopaedic Transition*

Translational research in orthopaedics is a fast-expanding area. Cellular and molecular research must be used properly in the therapeutic environment to really enhance people's health. In addition to bringing cutting-edge information to the forefront, this project will enable pioneers of orthopaedic translation to share and mutually improve skills.

3.10 *Translational Stem Cell Medicine*

To improve the therapeutic use of cellular and molecular biology of stem cell, Stem Cell Translational Medicine (SCTM) was STEM CELLS Translational Medicine will assist improve patient outcomes by bridging stem cell research and speeding translation of new lab findings into clinical trials. The research of stem cells has developed quickly, yet at an astonishing pace. This chapter aimed to provide a comprehensive list of stem cell types.

3.11 *Translational Proteomics*

Translation to decipher complicated disease processes, proteomics uses multidisciplinary methods. It emphasises fast distribution of new findings. It untangles complicated disease processes utilising multidisciplinary methods. Proteins are essential components of the physiological metabolic processes of cells and are essential components of living organisms. Most human diseases are caused by functional protein interaction dysregulation. In recent years, advancements in science and technology have enabled the study of protein interactions inside cells.

3.12 *Translational Neuroscience*

Developing novel treatments for neurodegenerative, neuropsychiatric, and developmental disorders is the goal of Translational Neuroscience. Brain anatomy and function research influence the development of novel treatments for neurological disorders. It is the process of bringing new treatments with quantifiable results to neurological illness patients. Conceived to help those at risk of or suffering from neurological illness convert the vast amount of fundamental neuroscience, neuropathogenesis, and neuroengineering knowledge into treatments and quantifiable benefits.

3.13 Molecule Therapy

The term "molecular therapy" refers to molecular alterations in cells. Vaccine development, preclinical target validation, clinical trials, and safety/efficacy studies are all subjects addressed in the study. Molecular targeted treatments utilise drugs to target particular molecules on the surface or within damaging cells. These chemicals help provide signals to cells to divide or grow. The medicines work by slowing the growth and spread of cancer cells while sparing healthy cells. Targeted treatments employ several medicines with varying effects. Scientists are trying customised treatments on both animals and people (clinical trials). But just a few targeted treatments have FDA clearance. The long-term efficacy and safety of targeted therapy are unknown.

4 Advances in Translational Bioinformatics

Translational bioinformatics involves in development of storage, analytical, and interpretative techniques that maximise the conversion of increasingly large biological and genetic data into proactive, preventative, predictive, and participatory health. The development of novel techniques for integrating biological and clinical data, as well as the evolution of clinical informatics methodology to include biological observations, are all part of translational bioinformatics research. Butte and Chen popularised the phrase when they published "Finding disease-related genomic research within an international repository: initial steps in translational bioinformatics" (Butte & Chen, 2006; Wilson et al., 2022). TBI has grown in prominence over the last decade, attracting a large professional community that has published results in high-impact journals and presented findings at national and international conferences.

Informatics was also emphasised during the annual AMIA Summit on Translational Bioinformatics, which was hosted in San Francisco. TBI has been studying how computational tools and techniques may be utilised to understand, analyse, and manage clinical data since its inception, advancing the discipline of systems biology in the process. To assist define this new subject, Altman organised a yearly review session at the AMIA annual conference (Altman, 2012). Altman emphasised the scope of translational research in health care, which encompasses illness treatment, prevention, and monitoring, as well as the discovery and evaluation of biomarkers and their application to areas like rare disorders (Burton & Underwood, 2007) or gene-disease correlations (Caufield et al., 2022; Denny et al., 2010). Translational bioinformatics, he claims, combines translational medicine with bioinformatics. Translational bioinformatics connects the two disciplines by developing algorithms to analyse fundamental molecular and cellular data in order to improve therapeutic outcomes. To enhance patient treatment and our knowledge of biology, TBI research combines data from molecular (DNA, RNA, proteins, small molecules, and lipids) and clinical entities (patients) (Altman, 2012).

5 Prospects of Translational Bioinformatics

Precision medicine relies on translational bioinformatics to support genetic, environmental, and clinical profiles of people, allowing genomic data to be turned into individualised therapy. The individuals working in this field tackle the scientific and statistical difficulties presented by genetic data in an unusual way. Translational bioinformatics includes clinical genomics, genomic medicine, pharmacogenomics, and genetic epidemiology. Precision medicine in translational biotechnology has ramifications in both clinical and therapeutic areas, such as drug discovery. Introducing new medicines would require the multimodal cooperation of clinical personnel, physicians, laboratory staff, biostatisticians, and bioinformaticians. Clinical genomics assists in the discovery of novel molecular biomarkers that are verified by clinically relevant genetic testing (Pagonet al., 2002).

Pharmacogenomics may be concerned with the genomic/clinical phenotypic connections with pharmacologically active drugs (Rubin et al., 2005). A few new techniques are being explored in clinical studies. Drugs for diseases like cancer, AIDS, cardiovascular disease, asthma, and Alzheimer's will be created utilising pharmacogenomics. Currently, pharmacogenomics studies factors that influence a drug's concentration reaching its targets. The use of gene expression from cell lines to predict patient drug response is currently controversial due to cell line variability. The accuracy of predicting in vivo medication response using a patient's baseline gene expression profile varied between 60 and 80%. A mix of inherited and nongenetic factors affects cancer growth and treatment resistance. Comparatively to public health and environmental registries, genetic epidemiology collects genome-based data (Little & Hawken, 2010). It is a process of converting fundamental research into a therapeutic environment. The complexity of the human physiology and the variety of ~~the~~ human population would be a limiting factor for real translation into clinical practice, but they would provide some inputs to future medical advances (Hopkins et al., 2021).

5.1 Translational Genomics in Clinical Care

While genetics examines single functioning genes, genomics analyses our whole DNA, recognising non-coding DNA's regulatory role and the intricate connections between many genes and the environment. Precision medicine seeks to promote health and treat illness more accurately by combining predictive, preventative, personalised, and interactive components. Recent advances in fundamental research have revealed new genetic variations and biomarkers. Over the coming decade, many anticipate significant advances in genetic testing and genome sequencing.

A genetics medicine service will depend on general practitioners to assist patients with diagnosis, treatment, and illness prevention. More doctors may explore adopting genetic testing and genome sequencing in the future years, with some expecting complete integration into routine medical treatment within 10 years. It may assist in diagnosis, prognosis, and therapy. Inhibitors of BRAF and Herceptin® (trastuzumab) are two examples. PARP drugs are more successful in treating ovarian cancer in individuals with BRCA gene mutations.

In addition to high-risk DNA variants, thorough genotyping may also help identify milk and gluten intolerances, as well as mucoviscidosis. Assembling genetic and HAS data may help uncover low penetrant variants. MFS is caused by mutations in fibrillin 1 (FBN1). MFS patients have significant clinical heterogeneity within and between families due to the disease's aetiology. TGFBR1, TGFB2, TGFBR2, MYLK1, MYH11, ACTA2, and SMAD3may assist identify individuals at risk for aortic aneurysms. Studying these high-risk individuals' aorta shapes may assist predict disease progression.

Translational genomics may potentially investigate gene networks of people with various disorders to learn more about their relationships. That is why almost half of Down's syndrome individuals exhibit ~~an~~ over-protection against heart problems linked with connective tissue. Recent research indicates FBN1 is increased in Down's Syndrome (usually down-regulated in MFS). The creation of genomic networks will help clarify the connections between various diseases. Understanding linked syndrome gene networks may lead to targeted gene therapy for illnesses.

Baby with long QT syndrome at Lucile Packard Children's Hospital Stanford. In this instance, the baby's heart stopped many times just after delivery. Gene mutations may cause Long QT syndrome. Finding the mutation's gene is essential to therapy. WGS revealed a previously characterised mutation, as well as variation of a new copy number in the TTN gene that targeted genotyping alone, would not have detected. It also took hours or days rather than weeks to get the response.

NGS whole-genome sequencing is vital in the study of complex diseases like cancer. The fact that drugs work differently in different patients with same cancer has long been a problem in cancer treatment. Drugs that target the unique signalling patterns of individual patients are currently being discovered using large-scale pharmacogenomics and personal genomics datasets. These include databases for the cancer cell lines. The NIH's Cancer Genome Atlas Project analysed the genomic profiles of over 10,000 people to discover new cancer subtypes. The variability of drug response is thought to be caused by patients with specific genomic aberrations. Large-scale datasets can be used to predict drug combinations, reposition of drugs, and delineate mechanisms of action. They are becoming increasingly important in drug development. Precision medicine can thus be tailored to individual patients' genomic profiles.

6 OMICs for Drug Repurposing and Discovery

Over the past 60 years, the cost of creating new medicines has increased significantly, with each new medication costing approximately 80 times more in 2010 than it did in 1960. The long FDA clearance procedure has also been widely addressed. The time it takes for a lead to be discovered and approved by the FDA is estimated to be 12 years. Consequently, a growing number of researchers are looking at high-throughput and computational drug development and repurposing. Recent efforts have concentrated on utilising omics data, particularly genomics, to identify novel therapeutic targets and develop new applications for existing medicines (drug repositioning).

Many new large-scale biological databases, in addition to the Human Genome Project, will aid researchers in better understanding illness origins and progressions. Biomolecular structural data can be found in the RCSB Protein Data Bank, as well as links to other biological resources including gene and pharmacology databases. Using mass spectrometry, ProteomicsDB, for example, identifies organ-specific proteins and translated long intergenic non-coding RNAs in the human proteome derived from tissues, cell lines, and bodily fluids.

The Human Metabolome Database currently contains approximately 40,000 annotated metabolites entries because of these advancements. It uses mass spectrometry and NMR spectroscopy to provide experimental and analytical metabolite concentration data. Databases are believed to aid in the transformation of clinical practice, specifically in metabolic disorders, such as coronary artery disease and diabetes. In fact, metabolomics is a rapidly expanding research area that encompasses both endogenous metabolites as well as chemical and biological substances that interact with the human body. Compounds from meals, medications, TCM, and the gut bacterial flora are being fingerprinted by researchers. These will eventually aid in our understanding of the host–pathogen-environment connection. These databases aid researchers in improve understanding the progression of complex diseases. Pattern mining and clustering can allow for the identification of new biomarkers. Clusters that are partitional (hard) or hierarchical (tree-like nested structure). These methods may be sped up by utilising multicore CPUs, GPUs, and FPGAs in parallel.

One way to describe the process of employing an FDA-approved medicine for a condition other than what it was originally approved for is "drug repurposing". Off-label use has mostly been motivated by chance in the past. Viagra, for instance, was originally designed to treat heart issues but is now used to treat erectile dysfunction. Early phase clinical trials are avoided by using a pre-approved medication, saving time and money.

Association research may lead to the discovery of new pharmaceutical targets. Sanseau et al. (2012) looked at prior GWAS results and discovered that 15.6% of them are already pharmaceutical targets (in comparison with 5.7% of the general genome). In 103,638 patients and controls, Okada et al. discovered 101 overall RA (rheumatoid arthritis) risk loci, 18 of 27 current RA therapeutic target genes, and three approved cancer medications that might be active against RA. Khatri et al. (2013) identified a comparable module of 11 genes in eight previous organ rejection datasets.

Scientists discovered two non-immunosuppressive medications that may be repurposed to regulate these genes in a mouse model. Drug-Gene Interaction Database (DGI) and PharmGKB are two resources that may assist in translating genomic research results into effective medications. The resources for TBI may be found in the Table 1.

Finally, an expanding set of computational and experimental techniques based on genetic and clinical data enables medication repositioning. An increasingly wide range of drug repositioning approaches may be used quickly and efficiently by combining translational bioinformatics, statistical methodologies, chemoinformatics, and experimental procedures. There are currently efficient techniques for systematic drug repositioning utilising huge libraries of biologically active molecules. Medicinal chemists and other translational experts can help reposition drugs.

7 Systems Biomedicine

Systems biology is a new multidisciplinary study that combines biology, mathematics, computer science, physics, and engineering. Most biological systems are too complicated for even the most sophisticated computer models to capture all system characteristics. A useful mode should be able to correctly comprehend the system under investigation and give trustworthy prediction results. To do this, a certain degree of abstraction may be needed, focusing on the system behaviours of interest while ignoring other aspects. Systems biology does not study individual genes or proteins one at a time, as has been the case for the last 30 years. Rather, it studies the interactions of all components in a biological system in action. With the goal of building formal algorithmic models for predicting process outcomes from component input, systems biomedicine is an emerging approach to biomedical research. Several important characteristics define the systems approach:

- Pursuit of quantitative and accurate data
- The datasets' comprehensiveness and completeness
- Willingness to define, quantify, and alter biological complexity
- Focus on component interconnection and networks
- Obsession with mathematically predicting outcomes.

Table 1 Resources for translational bioinformatics that are open to the public

Name	*URL*	*Comments*
PharmGKB	http://www.pharmgkb.org	PharmGKB is a curated resource for physicians and academics interested in the effect of genetic diversity on medication response
Phenotype Knowledgebase	http://phekb.org	Electronic phenotypic algorithms and their performance characteristics may be built, validated, and shared via an online collaborative repository
Pharmacogenomic Biomarkers in Drug Labels	http://www.fda.gov/drugs/scienceresearch/researchareas/pharmacogenetics/ucm083378.htm	Contains a list of FDA-approved medicines that include pharmacogenomic information on their labels
Clinical Pharmacogenetics Implementation Consortium (CPIC)	http://www.pharmgkb.org/page/cpic	Contain a list of the guidelines of CPIC for drug-gene interactions
NHGRI Catalog of GWAS studies	http://www.genome.gov/	Curated list of phenotypes and key results of GWAS studies
Catalog of PheWAS results	http://phewascatalog.org	Contains the catalogue of EHR PheWAS results
Drug-Gene Interaction database	http://dgidb.genome.wustl.edu	Data from 13 sources is used to provide a search interface for drug-gene interactions

(continued)

Table 1 (continued)

Name	*URL*	*Comments*
My Cancer Genome	http://www.mycancergenome.org	Contains data related to cancer mutations, treatments, and relevant clinical trials
ClinVar	http://www.ncbi.nlm.nih.gov/clinvar/	It contains current connections between human variants and phenotypes, as well as supporting data
SHARPn	http://phenotypeportal.org	SHARPn developed a compendium of computable phenotypic algorithms

Network studies have been done mostly on cell-based systems like immunology and cancer, or on homogenous tissues like the heart and liver. Using new bioinformatics methods, scientists discover microRNA-gene networks that are important in human inflammatory disorders and cancer.

7.1 Personalised Genomics

Personalised medicine treatment is crucial for patients to achieve the best possible outcomes while minimising adverse effects and high direct medical expenditures. Personalised medicine makes use of genetic and genomic testing. Whole-genome sequencing (WGS) examines the expression and interactions of all the genes in the human genome rather than just one. Rather of providing gene signature profiles based on the expression levels of individual component genes, genetic testing looks for single gene mutations or overexpression. Breast cancer genes BRCA-1 and -2, melanoma gene BRAF, and non-small cell lung cancer gene EGFR are examples. Breast, colon, and prostate cancer Oncotype DX tests, as well as the 70-gene test for breast cancer WGS, have grown easier, quicker, and less expensive since its beginnings. It's so easy that it might become a standard test for healthy people in primary care. However, interpreting WGS results may be difficult.

7.2 Genetic Testing

Genetic testing utilises human DNA, RNA, or proteins to look for gene variations, chromosomal abnormalities, or proteins linked to illnesses or disorders. An individual's risk of getting or passing on a genetic disease can be assessed using the results of a genetic test. There are already over 1,000 genetic testing available in the United States.

Most medical genetic tests result in changes in medical treatment, based on data from clinical trials and other medical practice.

- Identify a genetic illness.
- The risk of having a specific genetic disorder.
- Forecast the likelihood of adverse effects or an atypical reaction to a medication.
- Find a common disease's elevated risk.

Genomic tests must be accurate and clinically relevant to be useful. Analytic, clinical, and utility validity are required for genomic tests. In clinical validation, identifying and quantifying possible causes of biologic variation is a key aim. Patients gain from tests that are clinically useful. Using genetic data to drive patient treatment is changing our health care systems, but there are still obstacles. It examines how people are using personal genetic information for enhanced diagnoses, tumour profiling, and genomic risk assessments. Examples of how genomics is being used in everyday patient care are interwoven with the daily difficulties still confronting genomics integration into clinical practice, as well as methods being developed to overcome these hurdles.

7.3 Genomic Testing for Individuals to Inform Health Care

Several companies began offering direct-to-consumer genetic testing in 2008, providing information on genes for both health and leisure. Individuals can now acquire genetic testing without a doctor's prescription thanks to the availability of DTC genetic testing from companies like 23andMe (Mountain View, CA). People received test results as well as personalised information about their genetic origin, disease risk, and response to therapy.

DTC genetic testing raises several fascinating ethical, legal, and social considerations. For years, there was debate on whether or not these exams should be regulated. In November 2013, the FDA ordered 23andMe to stop marketing and offering health-related information services. The FDA designated these tests as medical devices, requiring formal testing and FDA approval for each test. The FDA accepted 23andMe's Bloom syndrome test application in February 2015 (http://www.fda.gov/NewsEvents/Newsroom/PressAnnouncements/UCM435003), and the firm stated in October 2015 that it would begin giving health information in the form of 36 gene carrier status. A 23andMe client can download their raw genetic data and assess the results using information from other websites, like Geneticgenie, Promethease, Interpretome, and openSNP.

Genetic testing may aid patients, according to a case presented at the 2014 American Neurological Association meeting. Alzheimer's disease runs in the family of one of the patients' mothers. She didn't know if she was a carrier or not. She didn't want to pass that mutation on to her children, though. Her doctors were able to choose embryos that did not carry the Alzheimer's disease gene mutation because to PGD testing. The patient was never tested, and the number of affected embryos (if any) was never revealed.

The ability to simultaneously investigate several genes or the entire genome brings up new possibilities in genomic medicine. Patients are challenging doctors about the applicability of genetic and genomic medicine to their own care, since new technologies promise better diagnoses and treatments. Others believe that incorporating genetics and genomics into routine clinical practice will be difficult.

7.4 Computational Health Informatics (CHI)

Computational health informatics (CHI) is a relatively recent area of study in and out of medicine. Information technology (IT) is an interdisciplinary that incorporates aspects of biological science as well as medicine. CHI studies how computers affect health care. Health informatics is the study of how to forecast a patient's health by gathering and analysing data from all areas of healthcare. Patient care is at the heart of health informatics research (HCO). The amount of medical and healthcare data available has grown dramatically during the last few years.

The fast development of new technologies has increased the amount of digital health data in recent years. The digital health data is massive and complicated in structure for conventional hardware and software. Some of the reasons why conventional systems fail to handle large datasets include:

- A wide range of organised and unstructured data including medical records, handwritten doctor notes, MRI, CT, and radiographic films.
- Healthcare informatics has noisy, heterogeneous, complicated, varied, longitudinal, and big datasets.
- Big data analytics and visualisation challenges.
- The requirement to increase data storage capacity, calculation capacity, and processing power.
- Improving patient care, data security, sharing, and lowering healthcare costs.

Thus, methods are required to handle and analyse such large, varied, and complicated information efficiently. Big data analytics, a common phrase for big and complicated datasets, is critical in handling enormous healthcare data and enhancing patient care. It also has the potential to save healthcare costs, improve treatments, customise medication, and assist clinicians in making individualised choices.

Transcriptome and proteome profiling have established how genetic information is expressed to determine phenotypes. The analysis of the connection between protein and mRNA levels shows the intricacy of gene expression regulation during dynamic transitions, particularly during steady-state and long-term state changes. It is important to note that the connection between protein levels and coding transcripts is significantly influenced by mRNA spatial and temporal fluctuation. In this section, we explain how protein concentrations may buffer mRNA variation (Liu et al., 2016). Proteomics-based mass spectrometry is a large and complicated field including numerous mass spectrometers, spectra, and search results. Quantitation in various scanning modes at various MS levels adds to the complexity. The most difficult task is quantifying post-translational modifications (PTM). Many various quantification methods have been published, some of which may be directly used for PTM quantification (Allmer, 2012). Translation regulates the proteome composition by converting mRNA coding sequences into polypeptide chains. For example, translatomics has revolutionised the study of cancer, bacterial stress response, and biological rhythmicity. The translational design may increase recombinant protein output by thousands of folds.

8 Related Work

Translational bioinformatics, systems biomedicine, clinical informatics, statistical genetics, and genomic medicine are all being enticed to play an increasingly important role in accelerating the translation of genome-scale studies to hypothesis-driven biological modelling, effective treatment, and tailored disease management or prevention. Over the last decade, technological improvements in high-throughput sequencing have resulted in a growing global capacity for easily creating nucleotide sequences. The 1000 Genomes Project was created in order to compile comprehensive genetic variation maps of individuals from distinct groups (1000 Genomes Project Consortium, 2015). For the integration of genetic data with clinical information, data from primary care, hospitals, outcomes, registries, and social care records should first be gathered using controlled clinical terminologies such as SNOMED Clinical Terms and the Human Phenotype Ontology (Köhler et al., 2017). The Global Alliance for Global

Health (GA4GH) is developing a shared framework of concepts for adoption with the goal of accelerating human health advancements, increasing efficiency, and lowering costs in order to ensure global interoperability of medical genetic data (Aronson & Rehm, 2015).

Genomic data enabled the Pan-Cancer Analysis of Whole Genomes (PCAWG) study, which is an international collaboration aimed at identifying common patterns of mutation in more than 2,800 cancer whole genomes from the International Cancer Genome Consortium. PCAWG proposes to induce genomic, transcriptomic, and epigenomic changes in 50 distinct tumour types and/or subtypes. This research has established the utility of merging data from several individuals' genomes, which can result in the identification of novel targets and disease mechanisms, as well as improved diagnostic and treatment outcomes for specific patients (Vamathevan & Birney, 2017).

Large-scale innovative research initiatives, such as the International Human Cell Atlas Initiative10, which aims to create comprehensive reference maps of all human cells, as well as devices, applications, wearables, and implantable technology, will add to this data explosion (Vamathevan & Birney, 2017).

8.1 The Ongoing Research Works Are Looking for:

- Developing novel approaches for analysing and merging large-scale datasets derived from transcriptomic, proteomic, genomic, and signalling pathways and networks analyses.
- Using machine learning and other modern computational tools to analyse large-scale biological data sets.
- Identifying and statistically evaluating molecular biomarkers for the diagnosis, prognosis, and classification of illnesses.
- Making use of cutting-edge bioinformatics technologies such as Blockchain, Internet of Things, and big data analytics.
- Demonstrates the value of translational bioinformatics approaches in viroinformatics, drug development, and repurposing.
- Includes translational healthcare and clinical uses of next-generation sequencing (NGS).
- Demonstrates translational medicine systems and their potential for healthcare improvement.

- Conducts research on medical image analysis, with a particular emphasis on CT scans and the detection of novel coronavirus infections.

8.2 Research Gap

Discovering, reusing, sharing, and analysing data are all dependent on metadata and data standards in Translational bioinformatics research. Several recent assessments imply that lack of acceptance of such standards is often related to difficulties in understanding, accessing, and using them. (Vamathevan & Birney, 2017).

The informatics literature currently lacks research on "active management" of data assets during the lifetime of a research project. Translational research produces and administers data "in the wild", meaning researchers rarely anticipate how the data can be used beyond the original intent. Leaving data management planning to the last minute diminishes the value of research data assets and restricts reuse possibilities.

8.3 Future Research Perspective

Patients with genetic disorders account for a sizable proportion of the world's population with unique healthcare demands. Recent estimates place the incidence of chromosomal diseases at 3.8 per 1000, single gene disorders at 20 per 1000, and multifactorial disorders at 646.4 per 1000. For example, in India, around 2% of new-borns have single gene or chromosomal problems, and 3% of couples have children with recurrent illnesses. Around 30% of major chronic illnesses and over 1,400 common single gene disorders have genetic variations (Chakrabarty et al., 2016). The laboratory service should be required to deliver cutting-edge genetic and genomic testing and analysis. For the reasons stated above, the spectrum of variations/mutations in faulty gene(s) is not uniform across the population, necessitating the use of modern genomic technology. Precision medicine aims to apply the right dose of the right treatment to the right patient at the right time. This requires integrating cutting-edge genetic technologies into healthcare system. Integrative perspectives on fundamental principles of genomic, proteomic, and computational biology help researchers as they apply qualitative systems methods to medical challenges (Tiberti et al., 2022).

9 Conclusion

The biological and healthcare industries are now covered by translational bioinformatics, which bridges the gap between bioinformatics and medical informatics. Current research is used to connect biological data with clinical informatics in translational bioinformatics. It requires analysing and sequencing an organism's whole genetic code, from genes to transcripts. The translational bioinformatics databases help biologists to learn about disease management and therapeutic development. A signalling pathway's operation in a cell can be studied using systems biology. A comprehensive view of a biological entity is created by combining genomic, proteomic, and bioinformatics data. Bioinformatics methods can be used to replicate the appearance of specific human diseases or healthy states. HGP discoveries are used in translational genomics to improve the diagnosis, prognosis, and therapy of complicated disorders. These innovations have revolutionised both healthcare and biomedical research. New tools and methodologies are needed to turn massive databases into usable knowledge.

References

1000 Genomes Project Consortium, Auton, A., Brooks, L. D., Durbin, R. M., Garrison, E. P., Kang, H. M., Korbel, J. O., Marchini, J. L., McCarthy, S., McVean, G. A., & Abecasis, G. R. (2015). A global reference for human genetic variation. *Nature, 526*(7571), 68–74. https://doi.org/10.1038/nature15393

Allmer, J. (2012). Existing bioinformatics tools for the quantitation of post-translational modifications. *Amino Acids, 42*(1), 129–138. https://doi.org/10.1007/s00726-010-0614-3

Altman, R. B. (2012). Translational bioinformatics: Linking the molecular world to the clinical world. *Clinical Pharmacology & Therapeutics, 91*(6), 994–1000. http://doi.wiley.com/10.1038/clpt.2012.49

Aronson, S. J., & Rehm, H. L. (2015). Building the foundation for genomics in precision medicine. *Nature, 526*(7573), 336–342. https://doi.org/10.1038/nature15816

Buccitelli, C., & Selbach, M. (2020). mRNAs, proteins and the emerging principles of gene expression control. *Nature Reviews. Genetics, 21*(10), 630–644. https://doi.org/10.1038/s41576-020-0258-4

Burton, J. L., & Underwood, J. (2007). Clinical, educational, and epidemiological value of autopsy. *Lancet (london, England), 369*(9571), 1471–1480. https://doi.org/10.1016/S0140-6736(07)60376-6

Butte, A. J., & Chen, R. (2006). Finding disease-related genomic experiments within an international repository: First steps in translational bioinformatics. *AMIA. Annual Symposium Proceedings. AMIA Symposium, 2006*, 106–110.

Caufield, J. H., Sigdel, D., Fu, J., Choi, H., Guevara-Gonzalez, V., Wang, D., & Ping, P. (2022). Cardiovascular informatics: Building a bridge to data harmony. *Cardiovascular Research, 118*(3), 732–745. https://doi.org/10.1093/cvr/cvab067

Chakrabarty, S., Kabekkodu, S. P., Brand, A., & Satyamoorthy, K. (2016). Perspectives on translational genomics and public health in India. *Public Health Genomics, 19*(2), 61–68. https://doi.org/10.1159/000442518

Chen, J., Zhang, D., Yan, W., Yang, D., & Shen, B. (2013). Translational bioinformatics for diagnostic and prognostic prediction of prostate cancer in the next-generation sequencing era. *BioMed Research International, 2013*, 901578. https://doi.org/10.1155/2013/901578

Chen, J., Qian, F., Yan, W., & Shen, B. (2013b). Translational biomedical informatics in the cloud: Present and future. *BioMed Research International*, 1–8. http://www.hindawi.com/journals/bmri/2013b/658925/

Denny, J. C., Ritchie, M. D., Basford, M. A., Pulley, J. M., Bastarache, L., Brown-Gentry, K., Wang, D., Masys, D. R., Roden, D. M., & Crawford, D. C. (2010). PheWAS: Demonstrating the feasibility of a phenome-wide scan to discover gene-disease associations. *Bioinformatics (oxford, England), 26*(9), 1205–1210. https://doi.org/10.1093/bioinformatics/btq126

Denny, J. C. (2014). Surveying recent themes in translational bioinformatics: Big data in EHRs, omics for drugs, and personal genomics. *IMIA Yearbook, 9*(1), 199–205. http://www.ncbi.nlm.nih.gov/pubmed/25123743.

Gochhait, S. et al. (2021). A machine learning solution for bed occupancy issue for smart healthcare sector. *Journal of Automatic Control and Computer Science* (6). Springer, ISSN: 0146–4116.

Hopkins, M. M., Ibanez, F., & Skingle, M. (2021). Supporting the vital role of boundary-spanning physician researchers in the advancement of medical innovation. *Future Healthcare Journal, 8*(2), e210–e217. https://doi.org/10.7861/fhj.2021-0091

Khatri, P., Roedder, S., Kimura, N., De Vusser, K., Morgan, A. A., Gong, Y., Fischbein, M. P., Robbins, R. C., Naesens, M., Butte, A. J., & Sarwal, M. M. (2013). A common rejection module (CRM) for acute rejection across multiple organs identifies novel therapeutics for organ transplantation. *The Journal of Experimental Medicine, 210*(11), 2205–2221. https://doi.org/10.1084/jem.20122709

Köhler, S., Vasilevsky, N. A., Engelstad, M., Foster, E., McMurry, J., Aymé, S., et al. (2017). The human phenotype ontology in 2017. *Nucleic Acids Research, 45*(D1), D865–D876.

Little, J., & Hawken, S. (2010). On track? Using the human genome epidemiology roadmap. *Public Health Genomics, 13*(4), 256–266. https://doi.org/10.1159/000279627

Liu, Y., Beyer, A., & Aebersold, R. (2016). On the dependency of cellular protein levels on mRNA abundance. *Cell, 165*(3), 535–550. https://doi.org/10.1016/j.cell.2016.03.014

Liu, Z. X., Cai, Y. D., Guo, X. J., Li, A., Li, T. T., Qiu, J. D., Ren, J., Shi, S. P., Song, J. N., Wang, M. H., Xie, L., Xue, Y., Zhang, Z. D., & Zhao, X. M. (2015). *Yi chuan = Hereditas, 37*(7), 621–634. https://doi.org/10.16288/j.yczz.15-003

Mohabatkar, H., Rabiei, P., & Alamdaran, M. (2017). New achievements in bioinformatics prediction of post translational modification of proteins. *Current Topics in Medicinal Chemistry, 17*(21), 2381–2392. https://doi.org/10.2174/1568026617666170328100908

Pagon, R. A., Tarczy-Hornoch, P., Baskin, P. K., Edwards, J. E., Covington, M. L., Espeseth, M., Beahler, C., Bird, T. D., Popovich, B., Nesbitt, C., Dolan, C., Marymee, K., Hanson, N. B., Neufeld-Kaiser, W., Grohs, G. M., Kicklighter, T., Abair, C., Malmin, A., Barclay, M., & Palepu, R. D. (2002). Genetests-geneclinics: Genetic testing information for a growing audience. *Human Mutation, 19*(5), 501–509. https://doi.org/10.1002/humu.10069

Ritchie, M. D., Moore, J. H., & Kim, J. H. (2020). Translational bioinformatics: Biobanks in the precision medicine era. *Pacific Symposium on Biocomputing, 25*, 743–747.

Rubin, D. L., Thorn, C. F., Klein, T. E., & Altman, R. B. (2005). A statistical approach to scanning the biomedical literature for pharmacogenetics knowledge. *Journal of the American Medical Informatics Association, 12*, 121–129. https://doi.org/10.1197/jamia.M1640

Sanseau, P., Agarwal, P., Barnes, M. R., Pastinen, T., Richards, J. B., Cardon, L. R., & Mooser, V. (2012). Use of genome-wide association studies for drug repositioning. *Nature Biotechnology, 30*(4), 317–320. https://doi.org/10.1038/nbt.2151

Vamathevan, J., & Birney, E. (2017). A review of recent advances in translational bioinformatics: Bridges from biology to medicine. *Yearbook of Medical Informatics, 26*, 178–187.

Yin, D., Ling, S., Wang, D., Dai, Y., Jiang, H., Zhou, X., Paludan, S. R., Hong, J., & Cai, Y. (2021). Targeting herpes simplex virus with CRISPR-Cas9 cures herpetic stromal keratitis in mice. *Nature Biotechnology, 39*(5), 567–577. https://doi.org/10.1038/s41587-020-00781-8

Wilson, A. C., Chiles, J., Ashish, S., Chanda, D., Kumar, P. L., Mobley, J. A., Neptune, E. R., Thannickal, V. J., & McDonald, M. N. (2022). Integrated bioinformatics analysis identifies established and novel TGFβ1-regulated genes modulated by anti-fibrotic drugs. *Scientific Reports, 12*(1), 3080. https://doi.org/10.1038/s41598-022-07151-1

Tiberti, M., Terkelsen, T., Degn, K., Beltrame, L., Cremers, T. C., da Piedade, I., Di Marco, M., Maiani, E., & Papaleo, E. (2022). MutateX: An automated pipeline for in silico saturation mutagenesis of protein structures and structural ensembles. *Briefings in bioinformatics*, bbac074. Advance online publication. https://doi.org/10.1093/bib/bbac074

CHAPTER 4

Application of Bioinformatics in Agricultural Pest Management: An Overview of the Evolving Technologies

Bhabesh Deka, Azariah Babu, and Uma Dutta

1 Introduction

In all living things of this universe, sustenance is a common threat. According to Malthus' renowned essay, food production grows in an arithmetic ratio while the population expands geometrically. Humans have understood from the beginning that they must cultivate crops and expand output to feed an ever-growing population. Since the beginning of agriculture, men have attempted to raise the uneven curve of food production, and this curve has tended to rise, barring famines, in the past.

B. Deka (✉) · A. Babu
Department of Entomology, North Bengal Regional Research and Development Centre, Nagrakata, West Bengal, India
e-mail: bhabesh.deka@gmail.com

U. Dutta
Department of Zoology, Cotton University, Guwahati, Assam, India
e-mail: uma.dutta@cottonuniversity.ac.in

S. Dutta and S. Gochhait (eds.), *Information Retrieval in Bioinformatics*,
https://doi.org/10.1007/978-981-19-6506-7_4

Despite the green revolution's tremendous success and the introduction of contemporary technology, many people continue to die of hunger, and a sizable percentage of the world's population remains hungry. To fight the ever-increasing threat of pests and diseases, we need to take a step back and look for fresh solutions (Delmer, 2005). Among all of these pests, arthropods, which are classified as insects, are the most destructive. Vector-borne diseases represent the biggest threat to all, even though we are exposed to a wide range of ailments throughout our lives. Insects are the primary disease carriers of many such diseases. Many mosquito-borne diseases continue to kill humans, including malaria, dengue fever, chikungunya, and filariasis. Numerous epidemics have occurred in tropical and subtropical countries as a result of vector-borne diseases in the last several years. Pest management science is an area of research that has evolved through time to address this issue. Pest control in the modern era constitutes a broad variety of disciplines requiring both basic and applied sciences to be successful. Integrated Pest Management (IPM) is the name given to this joint endeavour. As a result of severe bug infestation in certain countries such as the United States of America, from grains to cash crops, pests feed on a broad variety of goods and materials. According to the Indian situation, between 2005 and 2007, bud necrosis virus, American bollworm, *Helicoverpa armigera*, *Spodoptera litura*, red hairy caterpillar, and tobacco caterpillar, among other important predators, caused significant losses in sunflower, soybean, sugarcane, wheat, and apple. Because of this, not only does it contribute to the global food crisis, but it erodes the foundation of our rural economy as well. Several areas of computational biology are directly or indirectly engaged in pest control research. Many prominent bioinformatics applications have been examined and discussed in detail throughout this chapter.

2 An Assessment of the Current State of Pest Control

Biological and chemical techniques, together with the traditional knowledge that has been accumulated over the years, form the important components of an integrated pest control strategy. To fight introduced pests, conventional biological control programmes rely on predators, illnesses, and parasites native to the pest's origin. As a consequence of this approach, a large population is needed to avoid genetic diversity loss caused by population size bottlenecks. While an introduction is taking place, the effective population size of the originating population may tend to decrease. A small part of the population is restricted to guarantee any undesirable illnesses or pest species are eradicated. Due to the small population size, the process may take several generations, and it may result in inbreeding (Murty & Banerjee, 2011). In reality, only a small percentage of individuals released into a new range contribute to the following generation, reducing the size of the initial population. Thus, even if the initial collection is sufficiently enough to prevent severe bottlenecks, genetic diversity may decrease with time.

In the beginning, chemical pesticides were regarded as a miracle answer to the growing pest problem by farmers who suffered substantial losses due to insect-related crop damage. A surge in demand for chemical pesticides resulted as a consequence of farmers abandoning conventional pest control techniques by adopting this new strategy. Pesticides used incorrectly have led to unintended effects in the ecosystem. In recent years, concerns have been raised about biomagnification, or the build-up of xenobiotics that are more concentrated in the body than in the environment (Wiratno et al., 2007).

Risky microorganisms are not only bad for our major basic requirements but also damage the economical status of a country (Cavicchioli et al., 2019). Some of the world's poorer countries, which depend on agricultural growth, are bearing the brunt of this economic catastrophe. Several times attack by insects in India has resulted in similar conditions. States like Andhra Pradesh, Karnataka, Punjab, Uttar Pradesh, Maharashtra, and Haryana have been particularly hard-hit by recent year's drought. There are just a few prevalent pests, including stem borer, fruit and shoot borer, pod borer, and top shoot borer. It has been decided to take a few measures to guarantee that our harvest would survive. Central Plant Protection Stations (CPPSs) and Central Surveillance Stations (CSSs) were combined in 1992 to become Central Integrated Pest Management Centres (CIPMCs).

It has been a long time since India has made any significant attempts to solve these serious issues in tackling the pest to save crops. There are 31 CIPMC in 28 states and one union territory that have been created to tackle the problem. During the years 2007 and 2008, pest control incurred a large amount of money. Among the important areas where additional contributions were considered were the pests monitoring (8.16 million acres), the bio-control chemical field releases (1900 million acres), and the area coverage (7.00 million acres).

In addition to training and monitoring, this huge amount of money has been utilised to develop innovative ways of combating epidemics. Genetic (Handler & Beeman, 2003), molecular (Bhattacharyya & Bhattacharyya, 2006), and theoretical methods have lately been proposed and evaluated, and researchers across the world are working on increasingly sophisticated and complex ways. Computational biology will likely play a major role in this field shortly, if not already.

3 Information Technology: Scope in Biological Sciences

A branch of contemporary science that is multidisciplinary in nature, bioinformatics, or computational biology, uses information technology to address biological problems. Mathematical modelling and statistical techniques were employed for a long time to forecast a wide variety of biological features, but with the advent of computers, the area saw a significant transformation. In recent years, methods from the discipline of bioinformatics have become more popular in all areas of fundamental biology. Several of the most frequent applications include phylogenetic analysis, comparative sequencing, structure prediction, and discovery and validation of drug and pesticide targets (Banerjee et al., 2008; Munjal et al., 2018). There is a growing demand for data mining to shed light on previously undiscovered aspects of complicated biological processes as more data is made available to the general public (Abdel-latief et al., 2007; Munjal et al., 2018). An additional advantage of this cutting-edge technology is the ability to create databases with efficient data input and retrieval methods for data entry (Gochhait et al., 2021).

Database technology has emerged as a key component of information technology in recent years. Relational databases are extensively used in a broad variety of areas. Biology databases are more difficult to maintain than other databases because the data is so complex, and it requires a high degree of integration to handle and link a range of input files. Primary and secondary database systems have been developed and are now widely used in business and academia, as well as in the armed forces.

As a result, academics and researchers have access to hundreds of useful databases on the Internet, which are accessible for free or for a nominal fee, databases are becoming more useful in pest management because of the establishment of relevant and helpful pest information. It's worth mentioning that, one such database is Spodobase (Nègre et al., 2006; Punia et al., 2021), which has data on Spodoptera ESTs.

To produce the ESTs, five unique cDNA libraries derived from three different species of *S. frugiperda* were utilised. These contain fat body, midgut, and haemocytes samples from the *S. frugiperda*, together with the Sf9 cell line. Because they are vital for biological processes such as development, immunity, and insect-plant interactions, these tissues have been chosen. Current ESTs in the SPODOBASE database total 29,325 and are organised into redundancy-free groups (2294 clusters and 6103 singletons). More ESTs from *S. frugiperda* or other species may easily be added to the SPODOBASE as required or when new data becomes available. To locate information, text searches, pre-formatted queries, query help, and blast searches may all be utilised. On the other hand, GO-Slim is utilised to annotate both NCBI and *Bombyx mori* EST databases using the same GO-Slim vocabularies.

4 Integrated Pest Management (IPM) Using Information and Communication Technologies (ICT)

The goal of integrated pest management (IPM) is to keep pest numbers below the level that causes economic damage. IPM, the most successful system approach pest control technique, has reaped enormous benefits from information and communication technology (ICT) applications. FAO's IPM Programme Development Officer, Andrew Bartlett, has classified the many areas of IPM in which ICTs are being used (Xia, 2003). It can also help researchers analyse data and create scientific papers (Bartlett, 2002). These documents may now be emailed to farmers, extension workers, researchers, other scientists, and others, or posted on a public website. Scientists might utilise the Internet to great effect in countries like India, where Internet kiosks and smartphones are reaching rural regions. Several features of ICT make it appropriate for this purpose: ICTs are effective because they are quick, persistent, comprehensive, open, and interactive. It benefits from Internet forums and websites that allow users to ask questions. Farmers, government agencies, non-governmental organisations (NGOs), and other private organisations that advise and educate farmers to need a higher level of expertise and competence than previous extension systems. ICTs may be utilised to enhance decision-making and training in IPM (Bartlett, 2002). Identifying pest species, and their life cycle, scouting, monitoring, establishing ETL, choosing and deploying control measures, and assessing treatment efficacy may all be done using ICT. Initiatives like IPM CRSP, Africa IPM Link, and IPMNet have developed ICT information channels such as discussion forums, websites, email list servers, online databases, and multimedia CD-ROMs. A list of available environmentally friendly bio-control methods and a pest distribution model linked to weather monitoring systems are included in this decision support system.

5 Entomo-Informatics: A Prelude to the Concepts in Bioinformatics

Entomo-informatics is a scientific field that is essential in today's entomological studies. Data collected in sequencing facilities has led to a new field called bioinformatics. Many biologists are currently unaware of bioinformatics methodologies, tools, and databases, which may lead to missed opportunities or misunderstanding of data. Due to pesticide resistance, entomological research has increased relevance. It represents different entomological databases, and suitable their URL addresses. These databases include all proteins, biological reactions, and physiological processes. The NCBI (Benson et al., 2013), DDBJ (Tateno et al., 2002), and EMBL (Calabrese, 2019; Stoesser et al., 2002) are important shared platforms for storing biological data. The NCBI's Entrez platform is a powerful biological search engine that helps us find material. With so much research going on globally, an International Nucleotide Sequence Database Consortium (INSDC) was created to minimise data duplication in the previously stated genome databases (Arita et al., 2021; Cochrane et al., 2010; Stevens, 2018). NCBI's sequence databases receive genomic data from worldwide sequencing efforts and are the backbone of bioinformatics research. There are several nucleotide databases grouped under the nucleotide database category. The Entrez search engine returned substantial data for the term insecta. Researchers interested in individual genes will go farther, sequencing them and submitting them to the various repositories.

DroSpeGe is a search engine for Drosophila genomes. Drosophila genomes are made accessible for comparative investigations on this platform (Gilbert, 2007; Song et al., 2011). DroSpeGe is a tool for biologists to compare species differences and similarities, including genes, genome structure and evolution, and gene function relationships. It also includes the Blast search tool to find comparable sequences when new assemblies are added to genome maps. DroSpeGe includes a tool called BioMart for mining annotations and sequences. FlyBase is a Drosophila-only database with all the latest genome sequencing news (Marygold et al., 2013). Drosophila researchers use this database for gene annotations, phenotypic data, and expression data. GIFTS (Gene Interactions in the Fly) are a database of gene interactions in Drosophila pattern development. DRES has a collection of Drosophila EST sequences. The most researched creature has pioneered the way for scientists, as a model for geneticists, developmental and molecular biologists. The medically important pest and most dangerously considered are mosquitoes, which are vectors for serious diseases such as yellow fever (*A. aegypti*), Chagas disease (*R. prolixus*), malaria (*A. gambiae*), elephantiasis (*C. pipiens*), and typhus (the body louse *P. humanus*).

6 Pest Control Decision Support System

It is a set of algorithms that uses a corpus of information to attain expert-level competency in explaining problems (Banerjee et al., 2008). There is a need for a knowledge base and an inference engine in every expert system (ES). This is due to the wide range of applications of expert system technology to biological issues and it is very difficult to categorise them. There are several reasons to use an ES system in pest control monitoring support systems. When disaster assessment and early warning circumstances arise, ES may be utilised as a decision-making tool. It has been utilised to solve a wide range of agricultural issues using knowledge-based expert system technologies. *Nilaparvata lugens*, a brown planthopper, have been a major pest in Shanghai's rice-growing region since the early 1980s. A variety of weather conditions in tropical and temperate regions affect this pest's prevalence and dispersal rates. By draining nutrients from the phloem, BPH promotes leaf drying and tiller wilting. The importance of decision aids for evaluating the harm caused by this insect is essential for a successful battle against this bug and effective management methods. In this regard, an interactive WebGIS expert system for early warning of BPH disasters was created, including data on a variety of variables such as pest density and meteorological conditions. The system categorises disasters using if–then logic. The server-to-client

WebGIS browser offers early warning results and disaster management methods for BPH pandemic parameters given in the interactive WebGIS. PREDICT is yet another example of ES's full potential being fulfilled. EXSYS and INSIGHT2 + were used to create a portable expert system for assessing insect damage to red pine stands in Wisconsin. For foresters who are not familiar with forest pathology or entomology, PREDICT is a useful tool. There are 28 distinct pathogens, including mammalian, insect, and disease species in addition to two different types of abiotic damage, that PREDICT detects. Farmers and agribusiness professionals may utilise the Greenbug Expert System to manage green bugs and other cereal pests in winter wheat by using the Greenbug Expert System. On the Department of Entomology and Plant Pathology's website, one may find this expert system at http://entoplp.okstate.edu/greenbug/index.htm. Students will learn to identify and study common cereal aphids; identify and study natural enemies; determine treatment thresholds; and choose a pesticide from a list of available pesticides. An increasing number of industries are being affected by computers, smartphones, biotechnology products, the Internet, and so-called "smart" equipment. Private companies and public and/or private research are bringing these technologies to the agricultural sector. Pest control as part of integrated agriculture production is changing recognition to bioinformatics. GMOs, integrated systems, and precision agricultural equipment that provide site-specific IPM are three emerging technologies that are anticipated to have an impact on IPM adoption.

7 Integration of Agricultural Systems and Pest Biology

There is a dire need for new pest control techniques. Agricultural systems and pest biology expertise are required to practice sustainable agriculture. To assist in agricultural production management, models are computer-based, such as decision support systems (DSS) that integrate data and human knowledge. For example, Colorado State University is developing a Pest Management Decision Support System in the United States, and the work has already begun. To achieve this, a system was developed in cooperation with all other sectors of the food and fibre manufacturing industry. Producers, academics, and policymakers now have a better understanding of how to monitor and sustain an integrated model system. Fertility and pesticide inputs may be controlled on farms to maintain appropriate levels for groundwater and the ecosystem as a whole by utilising the many components of this system.

8 Identification of the Target Protein and Its Structural Analysis

When it comes to the time to increase crop yields and post-harvest quantities, chemical pesticides and biological pesticides are important. As of right now, pesticides are the most commonly utilised method of pest management systems. In the end, the target receptor or protein molecule determines whether a pesticide is effective or not (LópezPazos & Salamanca, 2008). The pesticide's effectiveness depends on its ability to target a certain insect is the first and foremost factor. Second, the pesticide must neither damage the crops or indirectly the consumers of such goods, and third, the target must be sufficiently reliable to eliminate the pest effectively and accurately. The enzymes chitinase, chitin synthase (Wang et al., 2019; Yang et al., 2021) are very effective targets for pesticides. As a result of this, target specificity should be addressed cautiously in the meantime, it is crucial to understand the structure and behaviour of the target protein molecule. For example, X-ray crystallography, nuclear magnetic resonance, and mass spectroscopy are all techniques used to identify the structure. They are also quite expensive and take a lot of time. These efforts have been made possible in large part using in-silico technologies. The simplest approach to get insight into a target structure is to use existing prediction algorithms. The ab-initio technique (Deka et al., 2021; Levine, 1991) and homology modelling methods (LópezPazos & Salamanca, 2008; Haddad et al., 2020) are useful and practical for comprehending a target molecule. Gibbons (1985) attributes the former to graph theory and the latter to homology. A common ancestor must exist for all species since all life began as a single cell. Structurally linked molecules will have both conserved and changing regions in common with each other. It is considered to be above the twilight zone if two sequences have more than 25% identity between them.

9 Target Pesticides and Their Virtual Screening

The pharmaceutical and biological sciences have grown more dependent on discovering new drugs to treat patients. Calculative methods are important in this very successful field and it is possible to detect pesticides using the same technique. A few important technologies have advanced rapidly in this area. When an HIV protease inhibitor was discovered, molecular modelling methods gained traction (Wlodawer & Vondrasek, 1998) and are two pillars that support the target-oriented screening method.

10 Computational Chemistry and Docking

Docking is an essential technique in computational chemistry for screening new pesticides. Among virtual screening methods, this method has attracted a lot of interest. For modelling receptor-ligand complex stability and its three-dimensional structure and is considered a wonderful tool. It is first necessary to investigate the conformational space of the ligands that bind to the target molecules, and then the set is scored based on anticipated binding affinity. There are a lot of different conformations of the ligand–protein complex that may be generated using this technique. An algorithm's docking type is determined by the search technique by (a) while docking, search for conformational space, (b) before docking, search for conformational space, or even (c) use incremental docking. The first set of algorithms optimises the shape and orientation of the tiny molecule inside the receptor-binding pocket. This linked optimisation problem is difficult to apply to large datasets because of its complexity. Monte Carlo, simulated-annealing, or genetic algorithms are frequently employed to deal with this problem (Lorenzen & Zhang, 2007).

11 Quantitative Structure–activity Relationship (QSAR)

Over the past four decades, QSAR has been widely utilised in agrochemistry, chemistry, toxicology, and pharmaceutical chemistry (Hansch et al., 2001; Kwon et al., 2019). As a result of rigorous testing and independent variable fine-tuning, the output of molecular and atom-based descriptors, as well as those produced by quantum chemical calculations and spectroscopy, has risen (Cho, 2005). It is now possible to screen a large number of substances under the same test conditions using high-throughput screening techniques. This reduces the risk of combining findings from different sources. It's time to go back to the basics. QSAR techniques are now multidimensional, ranging from 0 to

6 dimensions, depending on the application and validation methodology (Visco et al., 2002). Because it enables us to evaluate the importance of structural, geometric, thermodynamic, electrostatic, and other descriptors in connection to the biological activity of a chemical series, this sophisticated approach provides a technological complement to docking techniques. Both docking and QSAR are effective techniques for finding the most promising pesticide candidate chemical (da Silva Mesquita et al., 2020; Isyaku et al., 2020; Li et al, 2007). New pesticides with fewer negative effects are increasingly in demand. Through the use of virtual methods, we have made progress by decreasing the amount of time it takes to conduct experiments. These cost-effective, time-efficient, and accurate techniques form the basis of the contemporary approach to problem-solving point of view.

12 Pest Management: Role of Mathematical Modelling

Efforts have been made to model pest-affected areas spatiotemporally, as well as implement strategies to remove the danger. A wide range of advanced optimisation methods and artificial intelligence-based approaches are being utilised to solve different pest management problems (He et al., 2019; Karar et al., 2021). As a result of theoretical biological applications, pesticides have been manufactured more efficiently and more effectively. Artificial neural networks (ANNs) (Banerjee et al., 2008; Kujawa & Niedbała, 2021) and GA (Genetic Algorithms) are increasingly used in various optimisation fields. Researchers have shown that to maximise pesticide crystal protein (PCP) production on a large scale by using the *Bacillus thuringiensis* (Hofte & Whiteley, 1989); it is essential to regulate the concentrations of different processes variables, in particular, during fermentation. It has also being investigated if other pest control techniques based on the same bacteria might be used (Khasdan et al., 2007). A basic quadratic regression equation is used to connect the output variables to the process input variables in fermentation optimisation techniques. As a result of biological systems being so complicated, traditional methods are useless. Therefore, data-driven optimisation techniques are approaches, such as ANNs.

Various researches have shown that artificial intelligence techniques may be used to achieve different type's pest control goals. A model of the worldwide distribution of insect pests was created using self-organising maps (SOM) to detect geographical patterns and species assemblages (Gevrey et al., 2006). It was determined that each species posed a certain invasion risk using the SOM weights, allowing researchers to assess a species' potential to spread to a certain geographic area. Now, pest control uses more sophisticated mathematical modelling methods. As a result of their increased efficiency and accuracy, these methods are seldom employed in isolation. The Support Vector Machine (SVM) and heuristic techniques are two of these methods (Doran & Ray, 2014; Meyer et al., 2003). The quantitative structure-retention relationship (QSRR) was used to estimate the retention durations (RTs) of 110 different pesticides or toxicants (Li et al., 2007). There have been many complicated techniques created for assessing the probability of pest establishment following the introduction of a species into a region (Isman, 2019). Graphic techniques and computer-based decision-making tools like BIOSECURE (a biosecurity risk management tool for New Zealand's indigenous ecosystems) (Barker, 2010) were included. No comparison study has ever been done to further understand the best approach selected. As part of the investigation, a probabilistic estimate was made of the pest's likelihood of developing at each location based on a comparative study. The chosen method evaluated the likelihood of establishment before the pest was introduced. The consequence is a false awareness of pest threats before a real attack. As a consequence, many simulation methods aid in the creation of pesticide manufacturing, risk assessment for pest assault, and other jobs that are essential for developing an effective plan during a genuine attack or prevention.

13 Advanced Strategies

There is a lot of promise for microRNA-related techniques in a variety of emerging technologies (Kim & Nam, 2006). When it comes to digesting mRNAs, this amazing molecule is more precise. A pest's unwanted gene product will be much simpler to target, and pest population management will become significantly more precise in the foreseeable future. These techniques are used to study and produce pharmaceuticals (Ford, 2006). Apart from that, it may also be used to kill insects. The scientific community has not yet found this region. Computational biology has the potential to be very beneficial in miRNA screening (Pla et al., 2018). In the past, various microRNA screening techniques have been developed; however, this area still needs a great deal of research and specificity.

However, the application of information technology in pest management may be expanded. On the other hand, Benfenati et al. (2003) suggested log p values for 235 pesticides. Whether experimental or virtual, a single value (log P) should not be relied upon; instead, log P values should be calculated using many different techniques and the most consistent results should be used.

14 Key Issues and Major Challenges and Their Solutions

The advancement of Next Generation Sequencing (NGS) technology has transformed the field of genomics, allowing for the rapid, cost-effective, and accurate generation of sequencing data. These technologies are widely employed in a wide range of sectors, including pest control. However, because no one expected the rapid expansion of sequencing technology, computational biology is confronted with significant hurdles in data storage, administration, security, and analysis (Jayaram & Dhingra, 2010; Kushwaha et al., 2017). Since the invention of DNA sequencing in 1977, the technology has experienced a significant reduction in sequencing prices. Unfortunately, computers are not advancing at the same rate. Lower cost, and hence larger scale, of genome sequencing generates massive volumes of data, causing serious central processing unit (CPU) and storage issues. Since the handling of such massive data sets necessitates computing resources beyond the capabilities of a typical computer, there are two solutions. The conventional solution is to employ a computer cluster. This is not cheap, since it necessitates investments in hardware, software, physical storage space, and energy and cooling costs for the cluster. Moving calculations to the cloud is a newer, more cost-effective option. Cloud infrastructures are adaptable and dynamic, allowing users to adjust assigned resources up and down as needed. Because the cloud is managed by a third party, there is no need to be concerned about hardware or related expenditures (Jayaram & Dhingra, 2010; Kushwaha et al., 2017).

Among the most difficult challenges for bioinformaticians are the search for and use of publically available data, as well as coping with diverse formatting styles. A tremendous quantity of data is publicly available, yet using it might be difficult. After finding and downloading the data, it is critical to analyse it, ensure its quality and applicability, and avoid losing all of the information in the process. These are time-consuming procedures that frequently result in blunders. Researchers estimate that around 80% of their time is spent on data preparation and only 20% on real data analysis. This is due to a lack of defined file formats and uneven data formatting, which means that each new software generates a different data format. The idea is to automate these time-consuming and difficult data-grooming procedures, freeing up researchers' time to focus on data analysis. When data is submitted to a "format-free" data analysis platform, it "loses" the format and becomes a meaningful biological entity, with all objects of the same sort responding equally, regardless of underlying formatting variations. Other prevalent concerns include repeatability and organisational challenges, such as wrongly identified genes or a complete absence of data annotation. Keeping track of data provenance is critical, and specifics such as scripts or exact versions of data used must be meticulously documented so that the analysis can be replicated by someone else or in the future. Because reproducibility is essential for cumulative science, researchers should pay close attention to such issues (Kushwaha et al., 2017). Another issue that researchers worry about is losing track of their own computations, making it extremely difficult to duplicate the process on another set of data. Keeping track of data provenance is one approach to assure data repeatability. Noting down all scripts and parameters, on the other hand, takes time. As a result, automation would be a huge benefit here, saving researchers a lot of time and effort. Data curation and the usage of limited vocabulary are two topics in agricultural bioinformatics that require attention. Editing scientific data is vital for information transmission; consequently, well curated datasets must be regularly created through analysis by experts in the area, with the findings made available to the public. There is a need for collaborative research groups to exchange restricted vocabularies. The plant ontology (PO) and gene ontology (GO) consortia's work would aid in the consistent deployment of limited vocabulary databases. This will allow for the quick sharing of knowledge and information about agricultural pest management issues and techniques. It would be beneficial to link agricultural knowledge resources.

15 Future Prospects

Although computational biology has made a significant contribution to pest management via its numerous services, there is still more work to be done in this area. The number of specialised databases, such as Spodobase

(Nègre et al., 2006), is small in contrast to the hundreds of databases available for other subjects. Insect spatial–temporal modelling is another significant field, and while many new techniques are being developed, many more are still required which will take a lot of learning to track pest movements over time. However, the integrated decision support system at Colorado State University is a great example of real and successful use for computational biology or core information technology. India is still in its infancy when it comes to computer science and information technology. Still more work to be done in terms of development and strategy. Several agricultural research institutes and universities should take the lead in developing such a comprehensive approach to agricultural development. Numerous research institutions and universities in India are trying to adapt to the present technological change to provide the best pest control help possible. Indian authorities maintain tabs on mosquito species and the diseases they transmit (Murty et al., 2006). Using sequence features to group mosquito species has improved Anopheles classification, identification, and vectorial capability assessments (Banerjee et al., 2009). According to a new study, agricultural pest microRNA target prediction has been performed recently (Gualtieri et al., 2020; Jike et al., 2018; Moné et al., 2018; Singh & Nagaraju, 2008).

16 Conclusion

Future research in pest control should focus more on this important topic. Numerous significant uses of current multidisciplinary fields linked to computational biology to pest control are described in this article. The fact that the study is interdisciplinary lends credence to the idea that there is still a lot to be done. However, this is slowly changing. Due to the urgency and importance of the situation, several studies are presently being performed. Involvement from mathematicians and statisticians is growing and there is optimism that a global solution could be found soon if, they work together. These disciplines' combined efforts may one day help us find out a solution that does not need us to give up our food freely, or even a small part cheerfully. For this reason, we must position ourselves as the fittest to ensure our survival via the use of technical instruments in a situation where the survival of the fittest scenario occurs.

References

Abdel-latief, M. (2007). A family of chemoreceptors in *Tribolium castaneum* (Tenebrionidae: Coleoptera). *PLoS ONE, 2*(12), e1319.

Arita, M., Karsch-Mizrachi, I., & Cochrane, G. (2021). The international nucleotide sequence database collaboration. *Nucleic Acids Research, 49*(D1), 121–124. https://doi.org/10.1093/nar/gkaa967

Banerjee, A. K., Arora, N., & Murty, U. S. N. (2009). Clustering and classification of anopheline spacer sequences using self organizing maps. *The Internet Journal of Genomics and Proteomics, 4*(1).

Banerjee, A. K., Kiran, K., Murty, U. S. N., & Venkateswarlu, Ch. (2008). Classification and identification of mosquito species using artificial neural networks. *Computational Biology and Chemistry, 32*, 442–447.

Barker, K. (2010). Biosecure citizenship: Politicising symbiotic associations and the construction of biological threat. *Transactions of the Institute of British Geographers, 35*(3), 350–363. http://www.jstor.org/stable/40890992

Bartlett, A. (2002). *ICT and IPM, farmers, FAO and field schools: Bringing IPM to the grass roots in Asia* (pp. 8–9).

Benfenati, E., Gini, G., Piclin, N., Roncaglioni, A., & Vari, M. R. (2003). Predicting log P of pesticides using different software. *Chemosphere, 53*, 1155–1164.

Benson, D. A., Cavanaugh, M., Clark, K., Karsch-Mizrachi, I., Lipman, D. J., Ostell, J., & Sayers, E. W. (2013). *Genbank*. NAR D36–42.

Bhattacharyya, S., & Bhattacharya, D. K. (2006). Pest control through viral disease: Mathematical modeling and analysis. *Journal of Theoretical Biology, 238*, 177–197.

Calabrese, B. (2019). Standards and models for biological data: Common formats. In S. Ranganathan, M. Gribskov, K. Nakai, & C. Schönbach (Eds.), *Encyclopedia of bioinformatics and computational biology* (pp. 130–136). Academic Press. https://doi.org/10.1016/B978-0-12-809633-8.20418-4.

Cavicchioli, R., Ripple, W. J., Timmis, K. N., et al. (2019). Scientists' warning to humanity: Microorganisms and climate change. *Nature Reviews Microbiology, 17*, 569–586. https://doi.org/10.1038/s41579-019-0222-5

Cho, S. J. (2005). Hologram Quantitative Structure-Activity Relationship (HQSAR) study of mutagen X. *Bulletin of the Korean Chemical Society, 26*(1), 85–90.

Cochrane, G., Karsch-Mizrachi, I., & Nakamura, Y. (2010). The international nucleotide sequence database collaboration. *NAR*. https://doi.org/10.1093/nar/gkq1150

da Silva Mesquita, R., Kyrylchuk, A., Grafova, I., Kliukovskyi, D., Bezdudnyy, A., Rozhenko, A., Tadei, W. P., Leskela, M., & Grafov, A. (2020). Synthesis, molecular docking studies, and larvicidal activity evaluation of new fluorinated neonicotinoids against *Anopheles darlingi* larvae. *PLoS ONE, 15*(2), e0227811. https://doi.org/10.1371/journal.pone.0227811

Deka, B., Baruah, C., & Barthakur, M. (2021). In silico tertiary structure prediction and evolutionary analysis of two DNA-binding proteins (DBP-1 and DBP-2) from *Hyposidra talaca* nucleopolyhedrovirus (HytaNPV). *Biologia, 76*, 1075–1086. https://doi.org/10.2478/s11756-020-00665-x

Delmer, D. P. (2005). Agriculture in the developing world: Connecting innovations in plant research to downstream applications. *Proceedings of the National Academy Science, 102*(44), 15739–15746.

Doran, G., & Ray, S. (2014). A theoretical and empirical analysis of support vector machine methods for multiple-instance classification. *Machine Learning, 97*, 79–102. https://doi.org/10.1007/s10994-013-5429-5

Ford, L. P. (2006). Using synthetic miRNA mimics for diverting cell fate: A possibility of miRNA-based therapeutics? *Leukemia Research, 30*, 511–513.

Gevrey, M., Worner, S., Kasabov, N., Pitt, J., & Giraudel, J. L. (2006). Estimating risk of events using SOM models: A case study on invasive species establishment. *Ecological Modelling, 197*, 361–372.

Gibbons, A. (1985). *Algorithmic graph theory*. Cambridge University Press. http://pi.lib.uchicago.edu/1001/cat/bib/674902

Gilbert, D. G. (2007). DroSpeGe: Rapid access database for new Drosophila species genomes. *NAR, 35*, D480-485.

Gochhait, S. et al. (2021). The comparison of forward and backward neural network model—A study on the prediction of student grade. *Journal of WSEAS Transactions on Systems and Control, 6*. ISSN: 1991–8763, 422–429 (Scopus indexed)—Research Funded by UGC, Nepal.

Gualtieri, C., Leonetti, P., & Macovei, A. (2020). Plant miRNA cross-kingdom transfer targeting parasitic and mutualistic organisms as a tool to advance modern agriculture. *Frontiers in Plant Science, 11*, 930. https://doi.org/10.3389/fpls.2020.00930

Haddad, Y., Adam, V., & Heger, Z. (2020). Ten quick tips for homology modeling of high-resolution protein 3D structures. *PLOS Computational Biology, 16*(4), e1007449. https://doi.org/10.1371/journal.pcbi.1007449

Handler, A. M., & Beeman, R. W. (2003). United States department of agriculture-agricultural research service: Advances in the molecular genetic analysis of insects and their application to pest management. *Pest Management Science, 59*, 728–735.

Hansch, C., Kurup, A., Garg, R., & Gao, H. (2001). Chem-Bioinformatics and QSAR: A review of QSAR lacking positive hydrophobic terms. *Chemical Reviews, 101*(3), 619–672.

He, Y., Zeng, H., Fan, Y., Ji, S., & Wu, J. (2019). Application of deep learning in integrated pest management: A real-time system for detection and diagnosis of oilseed rape pests. *Mobile Information Systems, 4570808*, 1–14. https://doi.org/10.1155/2019/4570808

Hofte, H., & Whiteley, H. R. (1989). Insecticidal crystal proteins of *Bacillus thuringiensis*. *Microbiological Reviews, 53*, 242–255.

Isman, M. B. (2019). Challenges of pest management in the twenty first century: New tools and strategies to combat old and new Foes Alike. *Frontiers in Agronomy, 1*, 2. https://doi.org/10.3389/fagro.2019.00002

Isyaku, Y., Uzairu, A., Uba, S., Ibrahim, M. T., & Umar, A. B. (2020). QSAR, molecular docking, and design of novel 4-(*N*, N-diarylmethyl amines) Furan-2(5H)-one derivatives as insecticides against *Aphis craccivora*. *Bulletin of the National Research Centre, 44*(44), 1–11. https://doi.org/10.1186/s42269-020-00297-w

Jayaram, B., & Dhingra, P. (2010). *Bioinformatics for a better tomorrow*. Indian Institute of Technology.

Jike, W., Sablok, G., Bertorelle, G., Li, M., & Varotto, C. (2018). *In silico* identification and characterization of a diverse subset of conserved microRNAs in bioenergy crop *Arundo donax* L. *Science and Reports, 8*, 16667. https://doi.org/10.1038/s41598-018-34982-8

Karar, M. E., Alsunaydi, F., Albusaymi, S., & Alotaibi, S. (2021). A new mobile application of agricultural pests recognition using deep learning in cloud computing system. *Alexandria Engineering Journal, 60*(5), 4423–4432. https://doi.org/10.1016/j.aej.2021.03.009

Khasdan, V., Sapojnik, M., Zaritsky, A., Horowitz, A. R., Boussiba, S., Rippa, M., Manasherob, R., & Ben-Dov, E. (2007). Larvicidal activities against agricultural pests of transgenic Escherichia coli expressing combinations of four genes from Bacillus thuringiensis. *Archives of Microbiology, 188*, 643–653.

Kim, V. N., & Nam, J. W. (2006). Genomics of microRNA. *Trends in Genetics, 22*(3), 166–173.

Kujawa, S., & Niedbała, G. (2021). Artificial neural networks in agriculture. *Agriculture, 11*, 497. https://doi.org/10.3390/agriculture11060497

Kushwaha, U. K. S., Deo, I., Jaiswal, J. P., & Prasad, B. (2017). Role of bioinformatics in crop improvement. *Global Journal of Science Frontier Research, 17*(1), 1–13.

Kwon, S., Bae, H., Jo, J., & Yoon, S. (2019). Comprehensive ensemble in QSAR prediction for drug discovery. *BMC Bioinformatics, 20*(521), 1–12. https://doi.org/10.1186/s12859-019-3135-4

Levine, I. N. (1991). *Quantum chemistry*. Englewood Cliffs (pp. 455–544). ISBN 0-205-12770-3.

Li, X., Luan, F., Si, H., Hu, Z., & Liu, M. (2007). Prediction of retention times for a large set of pesticides or toxicants based on support vector machine and the heuristic method. *Toxicology Letters, 10*(175), 1–3, 136–144.

LópezPazos, S. A., & Cerón Salamanca, J. A. (2008). Mini review and hypothesis: Homology modelling of *Spodoptera litura* (Lepidoptera: Noctuidae) amino peptidase N receptor. *Revista De La Academia Colombiana De Ciencias Exactas, 32*(123), 139–144.

Lorenzen, S., & Zhang, Y. (2007). Monte Carlo refinement of rigid-body protein docking structures with backbone displacement and side-chain optimization. *Protein Science, 16*, 2716–2725.

Marygold, S. J., Leyland, P. C., Seal, R. L., Goodman, J. L., Thurmond, J. R., Strelets, V. B., & Wilson, R. J. (2013). FlyBase: Improvements to the bibliography. *NAR, 41*(D1), D751–D757.

Meyer, D., Leisch, F., & Hornik, K. (2003). The support vector machine under test. *Neurocomputing, 55*(1–2), 169–186.

Moné, Y., Nhim, S., Gimenez, S., Legeai, F., Seninet, I., Parrinello, H., negre, N., & d'Alencon, E. (2018). Characterization and expression profiling of microRNAs in response to plant feeding in two host-plant strains of the lepidopteran pest *Spodoptera frugiperda*. *BMC Genomics, 19*, 804. https://doi.org/10.1186/s12864-018-5119-6

Munjal, G., Hanmandlu, M., & Srivastava, S. (2018). Phylogenetics algorithms and applications. *Ambient Communications and Computer Systems: RACCCS-2018, 904*, 187–194. https://doi.org/10.1007/978-981-13-5934-7_17

Murty, U. S. N., & Banerjee, A. K. (2011). Bioinformatics with solutions in pest management science: An insight into the evolving technologies. In D. Reddy Vudem, N. R. Poduri, & V. R. Khareedu (Eds.), *Pests and pathogens: Management strategies* (pp. 521–542). BS Publications, CRC Press.

Murty, U. S. N., Rao, M. S., Arora, N., & Krishna, A. R. (2006). Database management system for the control of malaria in Arunachal Pradesh. *India Bio Information, 1*(6), 194–196.

Nègre, V., Hôtelier, T., Volkoff, A. N., Gimenez, S., Cousserans, F., Mita, K., Sabau, X., Rocher, J., Miguel, L. F., Emmanuelle, D., Audant, P., Sabourault, C., Bidegainberry, V., Hilliou, F., & Fournier, P. (2006). SPODOBASE: An EST database for the lepidopteran crop pest Spodoptera. *BMC Bioinformatics, 7*, 322.

Pla, A., Zhong, X., & Rayner, S. (2018). miRAW: A deep learning-based approach to predict microRNA targets by analyzing whole microRNA transcripts. *PLOS Computational Biology, 14*(7), e1006185. https://doi.org/10.1371/journal.pcbi.1006185

Punia, A., Chauhan, N. S., Singh, D., kesavan, A. K., Kair, S., & Sohal, S. K. (2021). Effect of gallic acid on the larvae of *Spodoptera litura* and its parasitoid *Bracon hebetor*. *Science and Reports, 11*, 531. https://doi.org/10.1038/s41598-020-80232-1

Singh, J., & Nagaraju, J. (2008). In silico prediction and characterization of microRNAs from red flour beetle (*Tribolium castaneum*). *Insect Molecular Biology, 17*(4), 427–436.

Song, X., Goicoechea, J. L., Ammiraju, J. S. S., Luo, M., He, R., Lin, J., Lee, S., Sisneros, N., Watts, T. A. D., Golser, K. W., Ashley, E., Collura, K., Braidotti, M., Yu, Y., Matzkin, L. M., McAllister, B. F., Markow, T. A., & Wing, R. A. (2011). The 19 genomes of drosophila: A BAC library resource for genus-wide and genome-scale comparative evolutionary research. *Genetics, 1,187*(4), 1023–1030. https://doi.org/10.1534/genetics.111.126540

Stevens, H. (2018). Globalizing genomics: The origins of the international nucleotide sequence database collaboration. *Journal of the History of Biology, 51*, 657–691. https://doi.org/10.1007/s10739-017-9490-y

Stoesser, G., Baker, W., van den Broek, A., Camon, E., Garcia-Pastor, M., Kanz, C., et al. (2002). The EMBL nucleotide sequence database. *NAR, 30*(1), 21–26.

Tateno, Y., Imanishi, T., Miyazaki, S., Fukami-Kobayashi, K., Saitou, N., Sugawara, H., et al. (2002). DNA Data Bank of Japan (DDBJ) for genome scale research in life science. *NAR, 30*(1), 27–30.

Visco, D. P., Jr., Pophalea, R. S., Rintoulb, M. D., & Faulon, J. L. (2002). Developing a methodology for an inverse quantitative structure-activity relationship using the signature molecular descriptor. *Journal of Molecular Graphics and Modelling, 20*(6), 429–438.

Wang, Z., Yang, H., Zhou, C., Yang, W. J., Jin, D. C., & Long, G. Y. (2019). Molecular cloning, expression, and functional analysis of the chitin synthase 1 gene and its two alternative splicing variants in the white-backed planthopper, *Sogatella furcifera* (Hemiptera: Delphacidae). *Science and Reports, 9*, 1087. https://doi.org/10.1038/s41598-018-37488-5

Wiratno, D. T., Paul, J. B., Ivonne, M. C. M. R., & Albertinka, J. M. (2007). A case study on Bangka Island, Indonesia on the habits and consequences of pesticide use in pepper plantations. *Environmental Toxicology, 10*, 405–414.

Wlodawer, A., & Vondrasek, J. (1998). Inhibitors of Hiv-1 protease: A major success of structure-assisted drug design. *Annual Review of Biophysics and Biomolecular Structure, 27*, 249–284.

Xia, Y. (2003). *The status of ICTs in integrated pest management*, (11), 1–2.

Yang, X., Xu, Y., Yin, Q., Zhang, H., Yin, H., Sun, Y., Ma, L., Zhou, D., & Shen, B. (2021). Physiological characterization of chitin synthase A responsible for the biosynthesis of cuticle chitin in Culex pipiens pallens (Diptera: Culicidae). *Parasites Vectors, 14*, 234. https://doi.org/10.1186/s13071-021-04741-2

CHAPTER 5

Application of Bioinformatics in Health Care and Medicine

P. Keerthana and Saikat Gochhait

1 Introduction

The field of biology has advanced into data research discipline and enormous wealth of data created is sequenced and available for synthesis of new information (Bayat, 2002). Thus, there is a need for careful storage, organization, synthesis and analysis of large data set for discovery of new knowledge. Hence, information technology and computation techniques are applied to biology to create a field called bioinformatics.

P. Keerthana (✉) · S. Gochhait
Symbiosis Institute of Digital and Telecom Management, Constituent of Symbiosis International Deemed University, Pune, Maharashtra, India
e-mail: keerthana.p2123@sidtm.edu.in

S. Gochhait
e-mail: saikat.gochhait@sidtm.edu.in

S. Dutta and S. Gochhait (eds.), *Information Retrieval in Bioinformatics*,
https://doi.org/10.1007/978-981-19-6506-7_5

Basic aims of bioinformatics are storing biological data in a database, developing tools to analyse data and interpreting meaningful results. The bioinformatics comprises computational and application bioinformatics. The computational bioinformatics deals with development of an application using computational work like, algorithm development, software development, database construction and curation for easy retrieval. Bioinformatics combines genetics with genomic technologies in discovering new clinical applications to estimate correlation between gene sequences and diseases, predicting protein structure from amino acid sequence and in discovering new drugs to customize treatment for individuals based on DNA sequence. The essentials of bioinformatics include development of software tool and algorithm, analysis and interpretation of biological data.

Bioinformatics tools are used for saving, retrieving, analysing and development of an efficient algorithm enabling measurement of sequence similarities (Table 1) (Mehmood et al., 2014; Rai et al., 2012). Bioinformatic applications include sequence analysis, molecular modelling, molecular dynamics, etc.

Bioinformatics has input from several areas of biotechnology and biomedical sciences. There are extensive variety of application of bio informatics in the arena of biotechnology in which the bioinformatic tools are used to compare the gene pair alignment that aid in identification of the functions of genes and genomes. It is also used in the study of molecular modelling, a method for analysing the three-dimensional structure of biological macromolecules, computer aided drug designing with the help of molecular docking, annotations, etc.

In genomics, bioinformatics is concerned with sequencing and analysis of genes (Collins et al., 2003). Structural genomics deals with genome structure identification, determination and characterization. Functional genomics focuses identification of genes, based on functions. Nutritional genomics is concerned with nutritional relevance to identified genes. Proteomics involves interaction proteomics to identify protein–protein association and expression proteomics for protein quantification.

Bioinformatics has various applications in the field of medicine ranging between research and molecular medicine, drug development, gene therapy and preventive medicines. With application of bioinformatics, new drug discoveries can be personalized to individual's genetic pattern. In

Table 1 Bioinformatic tools may be classified as follows:

Sequence analysis	*Homology and similarity*	*Protein function analysis*	*Structural analysis*
Align, CENSOR, SeWeR, Dna Block Aligner, Clustal W2, CpG Plot, MAFFT, MAUVE, HMMER, BioEdit, Vec Screen, ORF Finder, Pepinfo, SAPS, Transeq, etc	BLAST, FASTA, ENA, SSEARCH/GGSEARCH/GLSEARCH	CluSTr, Finger-PRINTScan, Inquisitor, InterProScan, Phobius, PPSearch, RADAR, Pratt, GLIMMER, Proteax	RASMOL, PyMOL, Qutemol, Ascalaph Designer, GROMACS, MDynaMix TINKER, NAMD, Jmol, Swiss-PdbViewer, MaxSprout, DaliLite, PDBeMotif, PDBeFold, Tempura

gene therapy, it can be used to manipulate the expression of an individual gene that have been adversely affected. Bioinformatics is also utilized in preventive medicine in combination with epidemiology. Drug development is one of the major applications of bio informatics along with computational tools, it aids in analysing the disease process and validate new and cost-effective drugs that target the cause of the disease. In cheminformatics and drug design, bioinformatics enable identification and structural modification of natural products and designing a drug with sought after properties.

Molecular genome, an application, helps in conducting DNA (Deoxyribonucleic Acid) sequencing and is useful in environmental clean-up, energy production, industrial processing and hazardous waste clean-up. Bio informatics in evolutionary studies helps to compare and determine the genome data of different species and characteristics.

Major application of bioinformatics in various sectors includes:

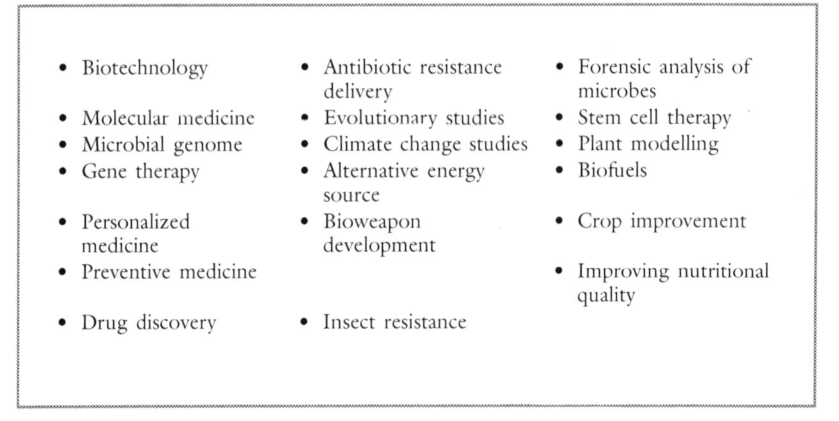

- Biotechnology
- Molecular medicine
- Microbial genome
- Gene therapy
- Personalized medicine
- Preventive medicine
- Drug discovery
- Antibiotic resistance delivery
- Evolutionary studies
- Climate change studies
- Alternative energy source
- Bioweapon development
- Insect resistance
- Forensic analysis of microbes
- Stem cell therapy
- Plant modelling
- Biofuels
- Crop improvement
- Improving nutritional quality

Limitations of bioinformatics are errors in sequence alignment, algorithm lacking capability to reflect the reality and exhaustive algorithm with slow rate computation. Hence, comparing results from different algorithms provide more accurate predictions.

1.1 *Major Challenges in Application of Bioinformatics*

The main challenges of bioinformatics include abundance volume of raw data, aggregate information and exponential growth of knowledge created by the study of genomes and its expressions. The increase in data volume poses a challenge in management of data, analysis of data and mining of data acquired that necessitates the need to adapt methods to address this knowledge base using formal approach like ontologies. The new knowledge thus acquired has to be in computational form for utilization. These challenges can be dealt with successfully in collaboration between the experts from various domains like computer scientists and biologists in dealing with the issues arising and enhance the application of knowledge developed in various fields (Lyon et al., 2004).

2 Bioinformatics and Secretome Analysis

Secretome is a collection of proteins secreted by the cells of an organism into the extracellular space and is essential in all the physiological, developmental and pathological processes. It includes enzymes, growth factors, cytokines, chemokine's, hormones, antibodies, interferon's and plays a vital role in cell proliferation, cellular immunity, cell differentiation, cellular communication and morphogenesis (Mukherjee & Mani, 2013).

As profiling of secretomes has become an indispensable area of research, an in-depth analysis and availability of computational tools help in systematic examination of secretome data. Thus, the bioinformatics helps in mapping and analysing protein sequences and aids in creation and visualization of 3D structure models and gene annotation.

Secretome analysis of unknown sequences is done by, Serial Analysis of Gene Expression (SAGE), protein microarray (antibody array and bead based array), signal sequence traps and mass spectrometry (Liquid chromatography tandem mass spectrometry (LC–MS/MS). It utilizes gel-based methods which include two-dimensional gel electrophoresis (2DE) and differential gel electrophoresis (DIGE) followed by mass spectrometry to identify secretory proteins. The gel-independent methods like iTRAQ (Isobaric tag for relative and absolute quantization), SILAC (Stable isotope labelling by amino acids in cell culture) and ICAT (Isotope-coded affinity tag) are often followed by LC–MS/MS or SELDITOF MS (Surface Enhanced Laser Desorption Ionization-Time-of-Flight mass spectrometry) (Jamesdaniel et al., 2009; Mukherjee & Mani, 2013).

The software's widely used in 2DE gel image analysis are Decyder, Progenesis, Dymension, GelScape, ImageMaster 2D, Melanie, Delta2D, PDQuest, Flicker, ProteomWeaver and LC/MS image analysis are—CPM, Decyder MS, MsInspect, OpenMS, OBI-Warp,OpenMS, ChAMS, PETAL, LCMSWARP, SpecArray, TOPP, MapQuant, Msight, etc. 2DE gel method is used to detect differential proteins associated with disease and treatment. An alternative to 2DE is the LC/MS proteomic outflow. Many software's are available for protein identification and quantitation like GutenTag, MASCOT, InsPecT, CPFP, Libra, Msquant, etc.

Secretome analysis for known sequences is done by RNA sequencing, DNA microarray and bioinformatics. Bioinformatics plays an important role in predicting secretory protein based on signal peptides. The secretory proteins with signal peptides (SP) predicted using computational analysis are SignalP and SecretomeP. The advantages of computational algorithm are identification of the secretory protein with SP or without SP. The disadvantages include absence of experimental validation, false positives and false negatives are frequently experienced.

Bioinformatic tools for prediction of secreted proteins are based on: Weight matrices (Predisi, SigCleave, SpScan, TMpred) Sequence alignment (Signal-BLAST (Basic Local Alignment Search Tool), WOLF PSORT) and Machine algorithm (MEMSAT, LipoP, ProtComp, PRED-Lipo, PRED-TAT, NClassG, SecretomeP, SignalP, SPOCTOPUS, TargetP, TATAFIND, TMHMM). Public repositories provide list of secretome profiles and links to external databases (e.g.: Interpro, PMAP, Pfam, Uniport, Swiss-Prot, PROSITE, Entrez Protein, PRIDE, FSD, OMIM, etc.). Development of bioinformatics tools and improvement in data bases facilitates experimental analysis of proteins thus beneficial for therapeutic purposes (Caccia et al., 2013).

3 Mass Spectrometry and Chemical Crosslinking Reagents

Mass spectrometry (MS) has advanced as an indispensable aspect in proteomics towards interpreting structural protein complexes (Matzinger & Mechtler, 2021). Application of mass spectrometry-based proteomics includes protein interaction networks, clinical proteomics, global proteomics, single cell proteomics, post-translational modifications, etc.

The two strategies widely used in protein identification and characterization are bottom -up and top-down approach. In bottom-up approach, the proteolytic digestion of proteins occurs before mass spectrum analysis. Intact protein is analysed by fragmentation in top-down approach and no digestion of protein is required and is time efficient (Chen et al., 2020).

Software tools are available for protein identification (MaxQuant, PEAKS, Mascot, TopPIC, PepHMM, SWPepNovo, EigenMS, SEQUEST, PECAN, PepNovo PROVALT, InsPecT, DirecTag, DBParser MassSieve, etc.) and protein quantification (iTracker, Libra, Label-free, OpenMS, Perseus, MaxQuant, IsobariQ, PVIEW, Skyline, Proteome Discoverer, XPRESS, PTMselect, etc.) (Petrotchenko & Borchers, 2010; Tran et al., 2015).

A number of downstream bioinformatics tools and databases are available (ProLoc-GO, PSEA-Quant, COVAIN, GNET2, IKAP, INGA, Pathview, KSEA, CRONOS, viper, STRING, SIGNOR, PANTHER, etc.) for proteomics analysis.

Crosslinking elucidates the structural information of protein. Intra-molecular or inter-molecular joining of two or more molecules via covalent bond with the use of crosslinking reagent is known as chemical crosslinking. Cross linking reagents are selected based on their chemical reactivity, cell membrane permeability, spacer arm length, chemical specificity, membrane permeability, cleavability and homo-bifunctional or hetero-bifunctional reactive groups. Cross linking reagents are very crucial and will affect the specificity and sensitivity of the analysis (Tran et al., 2015).

The crosslinking along with mass spectrometry is a powerful method in yielding structural information of protein and protein complexes (Back et al., 2003; Iacobucci et al., 2020).

There are many crosslinking mass spectrometry (CXMS) data analysis tools in use like: Xquest, StavroX/MeroX, Xisearch/XIFDR, ECL/ECL2, Mango, Kojak, pLink/pLink2, Proteome Discoverer/XlinkX SIM-XL, etc. Software tools for visualization to examine protein interaction networks are ProXL, XVis, CLMSVault, XiNET, Jwalk/MNXL, XlinkAnalyzer, Xwalk, TopoLink, etc. Information procured by crosslinking mass spectrometry helps in understanding of various important biological questions. It is also essential in assessing pathophysiological processes and aid in development of new drug discoveries. With the recent advances in bioinformatics tools, various computational issues of crosslinking mass spectrometry are addressed.

4 Bioinformatics and Software Product Line

Development in technology has led to a large repertoire of data being generated and it in turn necessitates the need of efficient storage, organization and retrieval of data. Data handling has become more streamlined with the arrival of bioinformatics. Many of the users of bioinformatics tools may not have appropriate training on software usage. As there is demand in different versions of a similar application in various researches, reuse of software is recommended. Hence, software product line (SPL) comes into play with development of common platform for a set of products committed to a particular domain. Software Product Line (SPL) utilizes the idea of "reuse". It is a group of systems created from a shared set of basic artefacts. In some cases, such as bioinformatics applications, it is desirable to use the SPL method to create a collection of linked software products (Costa et al., 2015).

The main concept of software product line is utilization of typical asset base in creation of an analogous set of products by core asset development, product development and management.

Three product line approaches are:

i. Proactive approach, suitable for organizations that can predict their product line requirement well in advance.
ii. Reactive approach is an extreme programming approach and is used when the requirements cannot be predicted beforehand.
iii. Extractive approach reuses existing software's for quick transitions without much reengineering.

The software product line constitutes multiple software with development of common features from common set of core assets (Northrop, 2002). The software assets include software requirements, architecture, design, documentation, prototypes, etc. The software product line comprises domain engineering concerned with core assets development and application engineering dealing with final products according to requirements of customers.

The Software Product line consist of two modules

1. Client Module
2. Core Module

4.1 Client Module

The Client module is mainly constructed using Web Interface. Web Interface is a technique that allows to interact with software or material on a distant server using web browser. The web server downloads the content of the web page, allowing us to interact with it using the browser. Scientific software's differ from existing one in the following aspects:

i. The method of creation of scientific software is not in traditional format.
ii. The software typically is developed by researchers and scientific experts themselves.
iii. Surveying and needs are both impeded since in an initial phase of study, they may not look obvious or occasionally even unknown.

4.2 Core Module

As the set of readings are ready, there will be two different analyses. If there is no availability of any genomic information before, it is necessary to assemble the reads into a genome or transcriptome (or) if genome is already available, it's possible to map our reads based on that genome. Although these two analyses appear similar, they are quite different. Mapping is the process of comparing each read to the reference genome and can get one or more sorts between every reading and genome.

The software product line architecture is essential to assure product features, organize development, control complexity and manage evolution. The effectiveness of SPL is based on investigation of the commonalities and variability among products. There is a need for significant level of commonality. The variations include time dimension and space dimension. Huge quantities of data are generated by technological advances and bioinformatic tools help in aggregating, analysing and interpreting these data's. Advantages of systematic reuse of product line utilization are enhancement of quality, improved productivity and decreased cost and time (Alvis, 2003). Bioinformatics serves as an important application domain involving management and analysis of mounting volume of varied data.

5 Bioinformatics and Protein Kinase

Healthcare system is confronted with many emerging morbid conditions where gene has an important role in deciding certain diseases (Gochhait et al., 2021; Rimal et al., 2021). Most of the diseases are linked to abnormal kinase activity like: cardiovascular diseases, cancer, bronchial asthma, hypertension diabetes mellitus, etc. Hence, investigation of kinase activity is indispensable in the field of drug designing and treatment of various diseases.

Protein kinase uses Adenosine triphosphate (ATP) to phosphorylate proteins. There are different groups of protein kinase which are further classified into families. Protein kinase and its substrates have important function in various cellular processes, and imperfections leads to diseases; thus, protein kinase plays vital role in drug discovery.

Different categories of protein kinase are: Serine/threonine protein kinases (adds the phosphate to serine or threonine), Tyrosine specific protein kinases (uses tyrosine as acceptor of phosphate), Histidine specific kinase, Aspartyl/glutamyl protein kinase and Tryptophan kinase.

Mechanism for substrate phosphorylation by a kinase starts with bonding of Adenosine triphosphate to the binding site of the kinase and followed by binding of the substrate. Phosphoryl transfer occurs, and substrate is released from the kinase, and this further leads to Adenosine diphosphate (ADP) release from the active site.

Two major factors determining protein kinase substrate complex are substrate recruitment and peptide specificity. The computational methods aimed at predicting substrate of protein kinases are based on the peptide specificity (Saunders et al., 2008). Prediction tools include NetwoKIN, KinasePhos, DISPHOS, PredPospho, Scansite, PPSP and NetPhos. The bioinformatic tools for analysis of posphopeptide data are simPhospho, Prophossi, PHOSIDA, PhosFox, etc.

An in-depth study of protein kinase using databases is essential to explore the regulatory mechanisms and to identify association between kinases and diseases. Application of bioinformatics is mainly utilized in analysing structure, function, regulatory mechanism, impact of kinases on diseases and efficient drug designing. A collection of kinase data bases are developed with the available protein kinase data to understand kinase pathways, they include: KKB (Kinase Knowledge Base) a Eidogen-Sertany's database of kinase chemical structure and biological activity, KinBase is a gene resource of protein kinase, Kinweb has compilation of protein kinases encoded in the human gene, KSD (Kinase Sequence Database) and KPD (Kinase Pathway Database) (Chen et al., 2015).

Bioinformatics thus helps in elucidating information of patient at the genetic level and combines genetics with genomic technologies in discovering efficient drug development with increased efficiency.

6 Bioinformatics and miRNA

Bioinformatics has transformed the field of biology; in fact, it has streamlined every aspect of biology from data storage to drug discovery. Like DNA, proteins and other biomolecules, the world of microRNAs (miRNA) had been integrated with the infinite potential of bioinformatics.

MicroRNAs are non-coding RNA molecule that is small and single stranded, regulating gene expression at translational or post-translational level. Thus miRNA act as key regulators of several cell processes such as cell proliferation, differentiation, metabolism and apoptosis (Lin et al., 2021). miRNA are capable of binding to their complementary mRNA's and bring about transitional repression either by blocking their access to protein ribosome complex or by degrading the mRNAs. miRNA's have now been emerged as a prospective biomarker for the identification of diseases like diabetes mellitus, cancer, viral infections, neurodegenerative diseases, etc. With an ever expanding application of miRNA's, more research are taking place in this field. Along with the research, next generation sequencing technology has revolutionized this discipline, furthering the building up of sequence information. Thus bioinformatics became indispensable in the world of miRNA's like any other (Huang et al., 2017).

Bioinformatics apart from its role in storage of enormous amount of data plays a pivotal role in miRNA nomenclature assignment, functional annotation, prediction of binding sites, miRNA target interaction, miRNA expression and biomarker information. A number of databases and algorithms have been developed towards this (Moore et al., 2015; Rasheed, 2017). These databases include,

- miRBase: creates a repository to assign stable names to newly found miRNA's in a consistent and stable manner.
- miRDB: serves as a repository for miRNA functional annotations and target predictions.
- miRWalk: clearly predicts the binding sites of miRNA associated with genes, pathways, cell lines, pathological conditions, etc.
- miRTarBase: presents a comprehensive picture of miRNA target interaction, to enable graphical visualization of interaction also contains many other features. It employs word cloud for this purpose and data source includes GEO, TCGA and CUP-seq.
- miRCancer: is a database designed to provide a comprehensive information on the expression of miRNA's in human cancer.
- doRiNA, SomamiR, etc.
- Early Detection Research Network (EDRN): Unlike other databases which stores sequence based information, EDRN is exclusively submitted to multiple types of biomarkers.

7 Clinical Bioinformatics and Cancer

Cancer is an abnormal proliferation of cells in an uncontrolled manner due to mutation, chromosomal rearrangement and error in molecular machinery. Bioinformatics has a definitive role in various areas of cancer research like single-cell analysis, cellular imaging, genome sequencing, biomarker discovery and analysis of transcriptome. Continuous research in the field of oncology identifies cancer biomarkers, which are a vital method in research and enable quick diagnosis, therapy and prognosis.

Bioinformatics in research is applied in diagnosing cancer, categorizing aetiologies, identifying various treatment methods and in assessment of prognosis. It also helps in understanding mechanism of initiation, progression and in identification of target. Bioinformatics facilitates complete analysis by connecting geographically available data of cancer patients. The data thus collected and organized are available in databases that allow data access and retrieval.

The international nucleotide sequence database collaboration (INSDC) covers DNA Data Bank of Japan (DDBJ), European Molecular Biology Laboratory (EMBL) and GenBank at National Centre for Biotechnology Information (NCBI). These repositories exchange data among themselves and are updated with the worldwide coverage and allow full access to all records. Multitude of studies on genome, transcriptome and proteome provides opportunity for investigators to analyse and integrate available data in cancer research. Many projects are developed to integrate the data types collected and combined from different sources. National Cancer Institute (NCI) has launched Cancer Genome Anatomy Project (CGAP) in cancer genetics, and Human Cancer Genome Project (HCGP) has developed millions of Expressed Sequence Tags (EST). These ESTs are integrated into International Database Cancer Gene Expression, which serves as a human cancer index (Stransky & Galante, 2010).

The NCI has also created a network related to cancer known as cancer biomedical informatics grid (caBIG) and facilitates in easy sharing of information to connect scientists and practitioners. It also builds tools to accumulate, analyse, integrate and disseminate information related to cancer and treatment. The data for cancer are available from International Cancer Genome Consortium (ICGC) in coordinating cancer genome studies and is concerned with the genomic changes in cancer globally, The Cancer Genome Atlas (TCGA) (Tomczak et al., 2015), National Cancer

Institute (NCI), National Human Genome Research Institute (NHGRI) and Catalogue of Somatic Mutations in Cancer (COSMIC) (Hinkson et al., 2017).

The available data can be used in bio simulation, early detection through identification of cancer biomarkers, risk analysis, identifying druggable targets and personalized medicine. Bioinformatics tools help in identifying mutations, molecular markers in different stages of cancer, profiles used in classification, diagnosis and predict clinical outcomes of cancer. The application of clinical bioinformatics in cancer is vital in creation of biobanks and incorporating them with clinical information for improved therapeutic outcome.

8 COVID-19 Implications and Sustainable Development Goals

The United Nation has developed an agenda for accomplishing sustainable development to provide peace and prosperity in the world. A global partnership strategy to attain sustainable development, seventeen goals were prioritized. These sustainable development goals (SDGS) aimed at removing poverty, reducing inequality, improvement in education and increased economic growth along with handling climate change and preserving forest and oceans.

The wide spread infection across the world due to corona virus disease has caused a huge adverse effect on the humankind. The global pandemic has caused healthcare crisis and also impacted global economy, thus hampering the achievement of sustainable development goals by 2030. The disruption caused by COVID-19 pandemic has led to delay in attainment of agenda set by the SDGS by reversing the progress earlier attained on hunger, poverty, education, health care, etc (Leal Filho et al., 2020; Min & Perucci, 2020; Ray et al., 2021).

The primary goal is to end poverty, but COVID-19 has led to increase in global poverty rate. There is an exacerbation of world hunger due to the global pandemic with worsening child malnutrition. The goal of ensuring healthy lives has reverse in progress with short life expectancy and disruption in healthcare system, leading to increase in morbidity and mortality during the pandemic. Provision of quality education is hampered by the COVID-19 pandemic situation, and the closure of educational institution has devastating effect on learning and well-being of students (Gochhait et al., 2021; Rimal et al., 2021). Gender equality is adversely affected with increased violence against women. In the pandemic scenario, people worldwide have to live without access to safe drinking water and proper hygienic service, as these are essential in containing the spread of COVID-19 infection.

Providing affordable and clean energy is essential to combat climate changes, which can be achieved by accelerating actions on modern renewable energy (Singh & Mishra, 2021). The decent work for all and sustainable economic growth is highly impacted by the pandemic that has led to the loss of jobs and is predicted for a substantial increase in unemployment of youth. The goal of building sustainable industrialization and fostering innovation has plunged as a result of COVID-19 crisis. Reducing inequality within and among countries is one of the goals of sustainable development that has a reverse progress and has in fact increased existing inequalities. The pandemic has worsened the quality of life and forced people to live in poor conditions and slowed down the goal of sustainable cities and communities. There is an urgent need to combat climate change, and its impact, the sustainability of oceans, seas and marine resources are under severe threat and need to be conserved as this serves as an important aspect of sustainable future.

Degradation of ecosystem has a deleterious effect on human well-being, and hence, it is important to promote terrestrial ecosystem and halt biodiversity loss. The global pandemic has exposed that there is a need for promoting peace and ensure justice for sustainable development. Trade tension, technological snag and dearth in financial resources have caused an unexpected blow to the global system and have retarded the global partnership for sustainable development.

There is an urgent need to invest in data, with innovation in acquiring data, processing, dissemination and modelling, which is a critical tool for containing the pandemic and help in risk assessment and prevention. The data on social and economic impact help to plan for recovery and remedial actions (Cannataro & Harrison, 2021). The unprecedented crisis and the severity of the consequence of pandemic is challenge for the government and communities.

References

Alvis, M. E. (2003). An introduction to software product line development, 26–37.

Back, J. W., de Jong, L., Muijsers, A. O., & de Koster, C. G. (2003). Chemical cross-linking and mass spectrometry for protein structural modeling. *Journal of Molecular Biology, 331*(2), 303–313.

Bayat, A. (2002). Science, medicine, and the future: Bioinformatics. *BMJ (Clinical Research Ed.), 324*(7344), 1018–1022.

Caccia, D., Dugo, M., Callari, M., & Bongarzone, I. (2013). Bioinformatics tools for secretome analysis. *Biochimica Et Biophysica Acta, 1834*(11), 2442–2453.

Cannataro, M., & Harrison, A. (2021). Bioinformatics helping to mitigate the impact of COVID-19—Editorial. *Briefings in Bioinformatics, 22*(2), 613–615.

Chen, C., Hou, J., Tanner, J. J., & Cheng, J. (2020). Bioinformatics methods for mass spectrometry-based proteomics data analysis. *International Journal of Molecular Sciences, 21*(8), 2873.

Chen, Q., Luo, H., Zhang, C., & Chen, Y. P. (2015). Bioinformatics in protein kinases regulatory network and drug discovery. *Mathematical Biosciences, 262*, 147–156.

Collins, F. S., Green, E. D., Guttmacher, A. E., & Guyer, M. S. (2003). A vision for the future of genomics research. *Nature, 422*(6934), 835–847.

Costa, G. C., Braga, R., David, J. M., & Campos, F. (2015). A scientific software product line for the bioinformatics domain. *Journal of Biomedical Informatics, 56*, 239–264.

Gochhait, S., Butt, S., De-La-Hoz-Franco, E., Shaheen, Q., Luis, D. M., Piñeres-Espitia, G., & Mercado-Polo, D. (2021a). A machine learning solution for bed occupancy issue for smart healthcare sector. *Journal of Automatic Control and Computer Science, 55*(6), 546–556. ISSN: 0146-4116.

Hinkson, I. V., Davidsen, T. M., Klemm, J. D., Kerlavage, A. R., Kibbe, W. A., & Chandramouliswaran, I. (2017). A comprehensive infrastructure for big data in cancer research: Accelerating cancer research and precision medicine. *Frontiers in Cell and Developmental Biology, 5*, 83.

Huang, J., Borchert, G. M., Dou, D., Huan, J., Lan, W., Tan, M., & Wu, B. (2017). Bioinformatics in microRNA research. In *Methods in molecular biology* (Vol. 1617). Humana Press.

Iacobucci, C., Gotze, M., & Sinz, A. (2020). Cross-linking/mass spectrometry to get a closer view on protein interaction networks. *Current Opinion in Biotechnology, 63*, 48–53.

Jamesdaniel, S., Salvi, R., & Coling, D. (2009). Auditory proteomics: Methods, accomplishments and challenges. *Brain Research, 1277*, 24–36.

Leal Filho, W., Brandli, L. L., Lange Salvia, A., Rayman-Bacchus, L., & Platje, J. (2020). COVID-19 and the UN sustainable development goals: Threat to solidarity or an opportunity? *Sustainability, 12*(13), 5343.

Lin, J., Zeng, J., Liu, S., Shen, X., Jiang, N., Wu, Y.S., Li, H., Wang, L., & Wu, J.-M. (2021). DMAG, a novel countermeasure for the treatment of thrombocytopenia. *Research Square.*

Lyon, J., Giuse, N. B., Williams, A., Koonce, T., & Walden, R. (2004). A model for training the new bioinformationist. *Journal of the Medical Library Association, 92*(2), 188–195.

Matzinger, M., & Mechtler, K. (2021). Cleavable cross-linkers and mass spectrometry for the ultimate task of profiling protein–protein interaction networks in vivo. *Journal of Proteome Research, 20*(1), 78–93.

Mehmood, M. A., Sehar, U., & Ahmad, N. (2014). Use of bioinformatics tools in different spheres of life sciences. *Journal of Data Mining in Genomics and Proteomics, 5*(2), 1–13.

Min, Y., & Perucci, F. (2020). Impact of COVID-19 on SDG progress (UN/DESA Policy Brief 18), 1–5.

Moore, A. C., Winkjer, J. S., & Tseng, T. T. (2015). Bioinformatics resources for DNA discovery. *Biomarker Insights, 10*(S4), 53–58.

Mukherjee, P., & Mani, S. (2013). Methodologies to decipher the cell secretome. *Biochimica Et Biophysica Acta, 1834*(11), 2226–2232.

Northrop, L. M. (2002). SEI's software product line tenets. *IEEE, 19*, 32–40.

Petrotchenko, E. V., & Borchers, C. H. (2010). Crosslinking combined with mass spectrometry for structural proteomics. *Mass Spectrometry Reviews, 29*(6), 862–876.

Rai, A., Bhati, J., & Lal, S. B. (2012). *Software tools and resources for bioinformatics research* (Vol. 1). New India Publishing Agency.

Rasheed, Z. (2017). Bioinformatics approach: A powerful tool for microRNA research. *International Journal of Health Sciences, 11*(3), 1–3.

Ray, M., Sable, M. N., Sarkar, S., & Hallur, V. (2021). Essential interpretations of bioinformatics in COVID-19 pandemic. *Meta Gene, 27*, 100844.

Rimal, Y., Gochhait, S., & Bisht, A. (2021b). Data interpretation and visualization of COVID-19 cases using R programming. *Informatics in Medicine Unlocked, 26*(6), 100705. ISSN: 0146-4116.

Saunders, N. F., Brinkworth, R. I., Huber, T., Kemp, B. E., & Kobe, B. (2008). Predikin and PredikinDB: A computational framework for the prediction of protein kinase peptide specificity and an associated database of phosphorylation sites. *BMC Bioinformatics, 9*, 245.

Singh, V., & Mishra, V. (2021). Environmental impacts of coronavirus disease 2019 (COVID-19). *Bioresource Technology Reports, 15*, 100744.

Stransky, B., & Galante, P. (2010). Application of bioinformatics in cancer research. In W. Cho (Ed.), *An omics perspective on cancer research.* Springer.

Tomczak, K., Czerwińska, P., & Wiznerowicz, M. (2015). The Cancer Genome Atlas (TCGA): An immeasurable source of knowledge. *Contemporary Oncology, 19*(1A), A68–A77.

Tran, B. Q., Goodlett, D. R., & Goo, Y. A. (2015). Advances in protein complex analysis by chemical cross-linking coupled with mass spectrometry (CXMS) and bioinformatics. *Biochimica et Biophysica Acta, 1864*(1), 123–129.

CHAPTER 6

Information Retrieval in Bioinformatics: State of the Art and Challenges

Sunita, Sunny Sharma, Vijay Rana, and Vivek Kumar

1 Introduction

The retrieval of biomedical literature is getting increasingly complex, necessitating the development of improved information retrieval systems. Unstructured resources, such as text documents, are scoured by Information Retrieval (IR) tools in massive data repositories, which are held on systems. Information depiction, storage, and groups are all phases of IR (Nadkarni, 2002), in which one of the most complex tasks in IR is formative which materials are appropriate to the users requirements and which are not. Users cannot perfectly propose search string in an exact

Sunita (✉) · V. Kumar
Department of Computer Science, Arni University, Kangra, India
e-mail: sunitamahajan2603@gmail.com

S. Sharma
Department of Computer Science, University of Jammu, Jammu, India

V. Rana
Department of Computer Science, GNA University, Punjab, India

S. Dutta and S. Gochhait (eds.), *Information Retrieval in Bioinformatics*,
https://doi.org/10.1007/978-981-19-6506-7_6

mode to get particular part of data from enormous data reserves under the current regime. Search results from basic information systems are of reduced worth. We plan another advancement to filtering searches to improve imply the user's data require grouping to work on the results of IR by utilizing distinctive inquiry extension strategies and produce a linear arrangement between them, where the linear arrangement was straightly between two development results all at once in our proposed framework for this section where the arrangements were linearly between two development results at time in our proposed system for this chapter. Query expansions, for example, discover synonyms and reweight original phrases to broaden the search query. They deliver substantially more targeted, specific explore outcomes than standard queries.

The rest of this chapter is facilitated as go after: "related work" segment gives a diagram of associated effort State of the Art|| Section discusses the terminologies used in information retrieval. Open Problems and Challenges|| Section outlines the existing problems and future scope thereof. Conclusion|| segment is ending, and it also handles on possibility of future work.

2 Literature

Because of the rapid growth of biological data, good IR systems are required offers on particular and meaningful responses to complex queries. One of the key challenges in information retrieval societies is query extension. Researchers have developed a variety of strategies for query expansion. Some methods stress the use of unstructured data (text documents) to determine expansion words, while others emphasize the use of structured data to determine expansion terms (Ontologies). Perez Aguera et al. (2010) compare various methods for query expansion in unstructured documents. The expansion term: Tanimoto, Dice, and Cosine coefficients to consider co-occurrence of terms in distinct papers. They also use Kullback Liebler Divergence to look at the allotment of development expressions in the peak position documents and the entire collected documents.

Described a work in Buscher et al. (2012) about how to choose words that are more pertinent to the query topic from feedback documents based on the placements of keywords in feedback documents. To solve this challenge in a coherent, probabilistic manner, they developed a position to importance model (PRM).

The conclusive of their experimentation on two major web-informational collections uncovers that the recommended PRM is very effective and hearty, results best in class pertinence models in both reports and passage-based input. Alipanah et al. (2011) introduced the necessary Expansion Terms (BET) and New Expansion Terms (NET) as new weighting techniques for ontology-driven query expansion index.

Rivas et al. (2014) developed query-expansion preprocessing strategies for fetching documents in different domains of biomedical literature from the Cystic Fibrosis corpus of MEDLINE documents. They performed tests to demonstrate the varying outcomes and benefits of utilizing stemming and stop-words in document and query segmentation.

Query expansion has been used in biological information retrieval research to improve retrieval performance. Abdou and Savoy (2008) used the SMART retrieval system to test the effectiveness of query expansion on MEDLINE resources. Literature of biomedical information retrieval, Xu et al. (2006) observed query-expansion approaches concerning with local analysis as well as worldwide examination and ontology. Matos et al. (2010) carry out an archive recovery and prioritization instrument that dependent on utilizes idea-arranged inquiry extension to discover reports that are connected to each other. For biological query expansion, Rivas et al. (2014) seem terms query-specific, corpus-specific, and phrases for language-specific. These investigations showed that by taking domain-specific factors into account, query expansion is able to improve efficiency of biomedical IR. According to these investigations, Dang et al. (2013) used a weighted-dependence model to assign weights to candidate concepts in order to increase retrieval effectiveness. Xu et al. (2006) used a blend of significance models to weight query development phrases for clinical search to discover patient cohorts.

3 State of the Art

3.1 Information Retrieval

Within a big document collection, information retrieval (Krallinger et al., 2017) is worried about distinguishing a subset of records whose content is usually applicable to a user's demand. The purpose of information retrieval given an immense data set of records and a particular data necessity—normally addressed as a question by the client—is to find the reports in the data set that meet the data need. Normally, the task must be completed correctly and quickly.

Index structures and Boolean queries: A Boolean inquiry (Dadheech et al., 2018; Du et al., 2020) is a basic and frequent approach to communicate an information requirement. A term (e.g., OLE1) or a Boolean term combination is provided by the user (e.g., OLE1 and lipid). The biomedical literature databases PubMed, as well as many other text search engines, use this query paradigm.

The vector model and similarity queries (Bordawekar & Shmueli, 2017), is based on give details in this section, is a widely used of Boolean query. The viewed documents are as (algebraic) vectors over terms in this configuration, as formally specify below. A search query, q, can contain a large number of terms and even an entire text. It is also represented as a vector and is observed as a body of content rather than just a set of search terms. The retrieval effort is reduced to looking for document vectors that are the most comparable to the search-query vector in the database. The several of documents' similarity measures have been developed and applied.

3.2 Text Categorization

Text categorization is a task that information retrieval systems frequently tackle (Tellez et al., 2018). This process is classifying text with category—tags from a collection of predefined categories. Categorization can be divided into two techniques. One option is knowledge engineering, in which the user by hand a set of rules to train expert knowledge about document categorization. The knowledge acquisition bottleneck is the key disadvantage of this strategy. A knowledge engineer must interview a domain expert and manually define rules. Any changes to the categories necessitate additional participation from the knowledge engineer.

3.3 IR in Biomedical-Informatics

The experimental methods based (Hersh, 2020) that allow for the investigation of genes and proteins from a whole genome are the initial step toward molecular comprehension of complex biological processes. While experimentation are designed and conceded, the ability to observe in the background of existing information and before hypotheses is critical for both informed planning and interpretation of outcomes (Rimal et al.,

2021). This kind of information is frequently found in the books. Individuals have had to follow during the literature, article by article, and gene by gene, in order to find it in the past.

4 Open Problems and Challenges

4.1 *Abundant Amount of Information*

As previously stated, researchers' consistent efforts have resulted in a massive growth in publications in the biological sciences. This volume of scientific literature necessitates more effort on the part of researchers, who are often tasked with staying current on all information relating to their chosen research topics. This endeavor is primarily driven by two factors: the ongoing rise in scientific outcomes and the absence of correspondence inside life science strengths. Creating fitting procedures, strategies, and apparatuses to help scientists in crafted by naturally separating the valuable material from the Web (especially from text sources) has turned into a significant worry in this situation.

The study subject of bioinformatics literature retrieval and mining is organically diverse, making the process of finding open problems and difficulties much more difficult. However, some specific difficulties, in our opinion, demand more attention from academics than others, resulting in considerable advances. Let us simply mention a few of them, without pretending to be exhaustive: I encoding/preprocessing strategies; (ii) the inherent difficulty of study of retrieval and mining challenges; (iii) principles and the need for more standardization; and (iv) judgment of existing tools.

4.2 *Encoding/Preprocessing Techniques*

The preprocessing approaches (Young et al., 2017) can be classified roughly along the subsequent dimensions: Natural-Language-Processing (NLP) (Jang et al., 2021), lexical techniques, and semantic approaches are the three types of techniques. Currently, NLP cannot ensure that viable solutions will be developed that can account for the almost limitless number of changes in how relevant information is "deployed" in text documents. Due to its enormous potential, this field may, nonetheless, become the key relevance in the near future. Lexical approaches that focus on finding important phrases that can characterize documents are usually easier to execute, regardless of whether they are framed in a frequency-based perspective or not. Actuality, they should only use as an initial stage, as merely lexical preprocessing, e.g., TFIDF (Dey et al., 2017) does not appear to be suited for ordinary literature retrieval and mining challenges. Semantic approaches are somewhere in the center between NLP and lexical techniques. When using semantic approaches, a common schema is to addition lexical information with supplementary knowledge, which can be collecting in a multiplicity of ways. Here are a not many models: I any text report can be planned to a current taxonomy/ontology, determined to recognize important ideas and joining them to the actual archive, to work with additional handling; (ii) explicit term disambiguation strategies (e.g., inactive semantic ordering, synset examination, or NER investigation) can be used, to improving the importance of candidate terms, to be used in further processing; Latent Semantic Indexing (LSI). Blynova (2019) is a technique that uses singular value decomposition to compute document and term similarities using a "soft" term matching criterion. Term and document similarity can be better assessed by the cosine of their vector expressions when phrases and documents are given as vectors of statistically independent components. With the introduction of synsets has grown increasingly prominent. In WordNet (Wang et al., 2020), English words are organized into synsets, which contain all synonyms connected to a single topic.

4.3 Fundamental of Literature Retrieval and Mining Problems

Aside from the problems associated with determining the "correct" encoding/preprocessing approach to use, some activities are intrinsically tough. Consider a generic open discovery process, framed in the LBD area that necessitates the selection of the hypothesis to be explored. Regardless of whether the connected work is coordinated by fitting heuristics pointed toward narrowing the scope of reasonable speculations, the fleeting intricacy of an open revelation measure stays high, necessitating the use of specific AI approaches and algorithms.

4.4 Principles and Necessities for Further Standardization

The field of life sciences is rapidly evolving. The purpose, a broad concurrence among scientists on how to describe biological concepts is becoming increasingly important. On the one hand, settling names, contractions, and abbreviations can be trying because of the way that the equivalent (or comparable) names, truncations, and abbreviations may be utilized to allude to different might be used to refer to various things. On the other hand, determining where a composite name begins and stops in a text might be tricky. These issues, in our opinion, are solely due to an absence of normal terminology and programming apparatuses. Luckily, the Unified Medical Language System (UMLS)—which unites various well-being and biomedical vocabularies and guidelines determined to empower interoperability between PC systems—has been a beneficial endeavor aimed at encouraging uniformity. The UMLS has three tools: a metathesaurus (which contains terms and codes from a variety of languages), a semantic network (which allows users to move between related categories and their relationships), and a specialist database (equipped with language processing tools). However, due to a lack of standardization, various other issues remain unsolved. Quite possibly the most troublesome hardships, as we would like to think, is the necessity for consequently consolidating writing and organic information bases. Bioinformaticians' obligations are particular from those of data set keepers. Standard apparatuses fit for removing messages and connections from the writing, just as helping data set custodians in finding pertinent material for comment, would extensively add to making the issue less serious or even missing for this situation. Other difficult issues have a close connection to the organization of scientific publishing

4.5 Measurement of Existing Tools

At present, scientists are trying to figure out how the proposed strategies might enhance access to existing materials. It has been possible to use new methodologies to explore and mine the literature through competitions. A recent study (Dogan et al., 2017; Drost & Paszkowski, 2017) did a critical examination of text-mining approaches in molecular biology and demonstrated the potential of this approach (BioCreAtIvE). Academics from around the world participate in this competition every two years to compare approaches for

i. detecting physiologically significant entities and their relationships to existing database entries.
ii. associations between entities and facts. By encouraging researchers to advance in their specialized areas of interest, further efforts in this regard could encourage the sharing of important knowledge and skills.

5 Conclusion

Bioinformatics text analysis aims to improve access to unstructured knowledge by easing searches, supplying auto-generated summaries, connecting publications with organized assets, displaying content more visually, and assisting analysts in the discovery of novel hypotheses. Research in bioinformatics text mining has developed over the past few years, from archive recovery to relationship extraction. There are now a number of tools that can be used to integrate literature analysis across a range of life science disciplines, and these tools are being developed at an increasing pace. As part of this study, we briefly discuss literature retrieval and mining in bioinformatics, text mining, and writing recovery.

We referenced certain issues worth extra investigation in the second piece of the paper, fully intent on creating bioinformatics writing recovery and mining techniques and frameworks. To summarize, the scientific community is effectively occupied with resolving many issues identified with writing recovery and mining, and a few arrangements have been introduced and executed. They will be that as it may, be generally pointless until mainstream researchers move toward all-inclusive guidelines for how existing information is published and disseminated with researchers, with an exacting focus on structure of scientific publications.

References

Abdou, S., & Savoy, J. (2008). Searching in Medline: Query expansion and manual indexing evaluation. *Information Processing & Management, 44*(2), 781–789.

Alipanah, N., Parveen, P., Khan, L., & Thuraisingham, B. (2011, July). Ontology-driven query expansion using map/reduce framework to facilitate federated queries. In *2011 IEEE International Conference on Web Services* (pp. 712–713). IEEE.

Blynova, N. (2019). Latent semantic indexing (LSI) and its impact on copywriting. *Communications and Communicative Technologies,* (19), 4–12.

Bordawekar, R., & Shmueli, O. (2017, May). Using word embedding to enable semantic queries in relational databases. In *Proceedings of the 1st Workshop on Data Management for End-to-End Machine Learning* (pp. 1–4).

Buscher, G., Dengel, A., Biedert, R., & Elst, L. V. (2012). Attentive documents: Eye tracking as implicit feedback for information retrieval and beyond. *ACM Transactions on Interactive Intelligent Systems (TiiS), 1*(2), 1–30.

Dadheech, P., Goyal, D., Srivastava, S., & Choudhary, C. M. (2018). An efficient approach for big data processing using spatial Boolean queries. *Journal of Statistics and Management Systems, 21*(4), 583–591.

Dang, V., Bendersky, M., & Croft, W. B. (2013, March). Two-stage learning to rank for information retrieval. In *European Conference on Information Retrieval* (pp. 423–434). Springer.

Dey, A., Jenamani, M., & Thakkar, J. J. (2017, December). Lexical TF-IDF: An n-gram feature space for cross-domain classification of sentiment reviews. In *International Conference on Pattern Recognition and Machine Intelligence* (pp. 380–386). Springer.

Dogan, R. I., Chatr-aryamontri, A., Kim, S., Wei, C. H., Peng, Y., Comeau, D. C., & Lu, Z. (2017, August). BioCreative VI precision medicine track: Creating a training corpus for mining protein–protein interactions affected by mutations. In *BioNLP 2017* (pp. 171–175).

Drost, H. G., & Paszkowski, J. (2017). Biomartr: Genomic data retrieval with R. *Bioinformatics, 33*(8), 1216–1217.

Du, L., Li, K., Liu, Q., Wu, Z., & Zhang, S. (2020). Dynamic multi-client searchable symmetric encryption with support for Boolean queries. *Information Sciences, 506*, 234–257.

Hersh, W. (2020). Information retrieval: A biomedical and health perspective. *Health Informatics*. https://doi.org/10.1007/978-3-030-47686-1

Jang, H., Jeong, Y., & Yoon, B. (2021). TechWord: Development of a technology lexical database for structuring textual technology information based on natural language processing. *Expert Systems with Applications, 164*, 114042.

Krallinger, M., Rabal, O., Lourenco, A., Oyarzabal, J., & Valencia, A. (2017). Information retrieval and text mining technologies for chemistry. *Chemical Reviews, 117*(12), 7673–7761.

Matos, S., Arrais, J. P., Maia-Rodrigues, J., & Oliveira, J. L. (2010). Concept-based query expansion for retrieving gene related publications from MEDLINE. *BMC Bioinformatics, 11*(1), 1–9.

Nadkarni, P. M. (2002). An introduction to information retrieval: Applications in genomics. *The Pharmacogenomics Journal, 2*(2), 96–102.

Pérez-Agüera, J. R., Arroyo, J., Greenberg, J., Iglesias, J. P., & Fresno, V. (2010, April). Using BM25F for semantic search. In *Proceedings of the 3rd International Semantic Search Workshop* (pp. 1–8).

Rimal, Y., Gochhait, S., & Bisht, A. (2021). Data interpretation and visualization of COVID-19 cases using R programming. *Informatics in Medicine Unlocked, 26* (6), 100705. Elsevier, ISSN: 0146-4116.

Rivas, A. R., Iglesias, E. L., & Borrajo, L. (2014). Study of query expansion techniques and their application in the biomedical information retrieval. *The Scientific World Journal,* (1), 1–10.

Tellez, E. S., Moctezuma, D., Miranda-Jiménez, S., & Graff, M. (2018). An automated text categorization framework based on hyperparameter optimization. *Knowledge-Based Systems, 149*, 110–123.

Wang, Y., Wang, M., & Fujita, H. (2020). Word sense disambiguation: A comprehensive knowledge exploitation framework. *Knowledge-Based Systems, 190*, 105030.

Xu, X., Zhu, W., Zhang, X., Hu, X., & Song, I. Y. (2006, October). A comparison of local analysis, global analysis and ontology-based query expansion strategies for bio-medical literature search. In *2006 IEEE International Conference on Systems, Man and Cybernetics* (Vol. 4, pp. 3441–3446). IEEE.

Young, N. E., Anderson, R. S., Chignell, S. M., Vorster, A. G., Lawrence, R., & Evangelista, P. H. (2017). A survival guide to Landsat preprocessing. *Ecology, 98*(4), 920–932.

CHAPTER 7

Hybrid Support Vector Machine with Grey Wolf Optimization for Classifying Multivariate Data

M. Revathi and D. Ramyachitra

1 Introduction

The categorization process consists of the most important aspects regarding the datamining idea. The classification mechanism has been discovered to occur frequently in everyday life. For example, in a railway station, tickets are distributed and classified based on the class required, in a hospital, patients are classified based on the nature of their disease and their risk factors (low, medium, and high), in a school, teachers classify students' performance based on the grade received (first class, second class, third class, and fail), and in mobile technologies, (Orriol-sPuig et al., 2009) the basic goal of multivariate data classification is

M. Revathi (✉)
Department of Biotechnology, Bharathiar University, Coimbatore, India
e-mail: aarya44v44@gmail.com

D. Ramyachitra
Department of Computer Science, Bharathiar University, Coimbatore, India

S. Dutta and S. Gochhait (eds.), *Information Retrieval in Bioinformatics*,
https://doi.org/10.1007/978-981-19-6506-7_7

to compute a decision boundary, also known as a class or plane separation boundary that divides the input data space into one or more classes. The nature of the problem domain determines whether a decision boundary or a class dividing border is a simple straight line, a sophisticated linear or non-linear representational form. Originally, the data used on a day-to-day basis was simple in nature and was grouped together based on the data space's specific qualities. As a result, obtaining a judgement boundary class is easier when the data groups are non-overlapping (Beckett Claire et al., 2017). New models for healthcare services and other datasets are being incorporated in present medical practises and industries to minimise the impact of various ailments and rising healthcare expenses. In the creation of computing environments in smart hospitals, classification and other data mining methods play a key role. Classification algorithms, for example, are quite effective in the classification of patient activity. In the past, computers were used to construct knowledge-based decision support systems that used domain knowledge from medical specialists and manually transferred that knowledge into rules or computer algorithms. Furthermore, clinical judgments were frequently made largely on the doctor's intuition and experience rather than the database's knowledge-rich data (Kalimuthu, Sivanantham, 2021). This practice resulted in unfavourable biases and errors. Excessive medical costs are also a possibility, which may have an impact on the quality of service supplied to patients. A medical diagnostic might offer incorrect results in a variety of reasons, including the fault of the doctor or hospital staff. This procedure takes time and is highly reliant on the subjective opinions of medical experts. The present paper introduces a grey wolf optimization technique according to support vector machines (SVM) to gather knowledge automatically from examples or raw data to address this challenge (Liu et al., 2021). Explain the general classifier model in Fig. 1 to perform data classification. The design of a data classifier model, as shown in Fig. 1, begins with the collection of raw data for any real-world situation, followed by the division of the obtained data into training and testing datasets (Ros, German et al., 2016). When the categories of classes under consideration overlap, the primary job is to find the discriminant function's optimal point, which reduces the number of misclassifications in the given data while also minimising the chance of misclassification of the unavailable data. The primary difficulties that underpin data classification are generalisation and approximation, based on this aspect (Kalimuthu, Sivanantham et al., 2021).

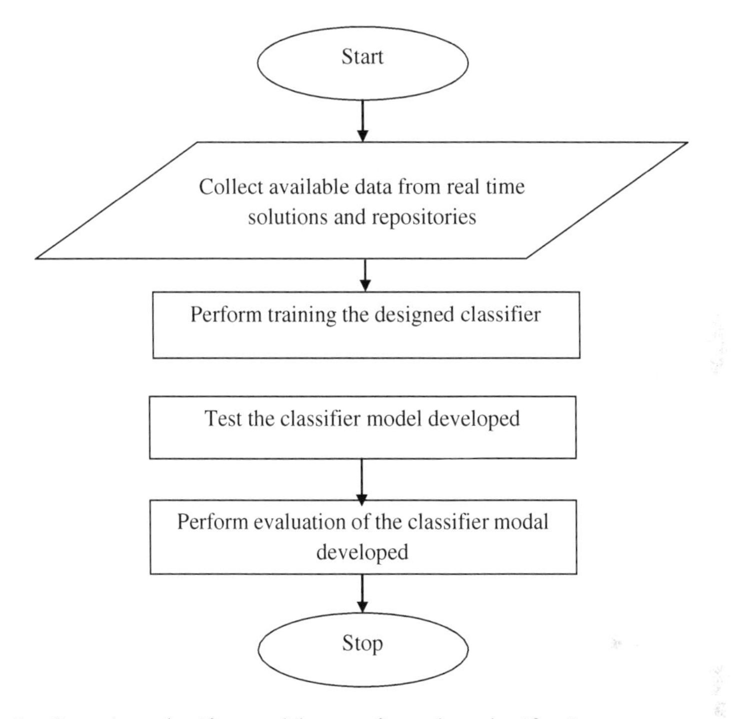

Fig. 1 Steps in a classifier model to perform data classification

To update parameters and learn the classifier structure, the classifier model uses a percentage of training data from the original datasets and prior knowledge of the issue area during the training phase. The trained classifier is evaluated in terms of the percentage of test data by offering a classification judgement for the considered input pattern during the test phase. The comparison architecture of existing and hybrid SVM-GWO algorithms is shown in Fig. 2. To test the suggested technique, you'll need a standard multivariate data set. An acceptable UCI data set containing heart disease diagnosis called "Cleveland, vowel, glass, shuttle, and yeast data" is used to evaluate the algorithms under consideration. Only a few entries in the shuttle-2 vs 5 data sets had missing values: yeast-0-3-5-9 vs 7–8, vowel0, cleveland-0 vs 4, and glass-0-1-4-6 vs 2.During data cleaning, all such records were eliminated. As a result, the total number of records is 3365, 1805, 545, 297, and 195.

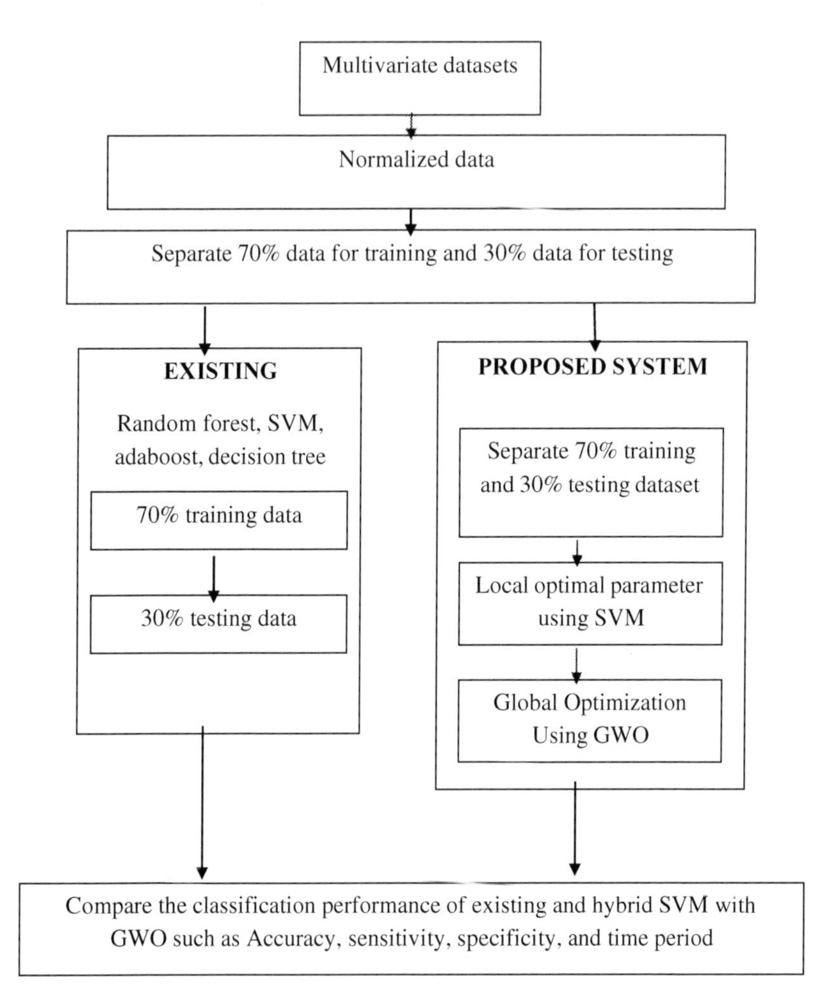

Fig. 2 Comparison architecture of existing and SVM–GWO

When the input data is normalised, almost all classification methods improve classification accuracy. The data was normalised after loading the CAD data by dividing each attribute value by the maximum value of that particular characteristic in this suggested evaluation method. This approach converts the data to values between 0 and 1, which are suitable for almost any classification method. The reminder for this paper is organised as follows. Section 2 discusses several data classification methods and investigations. Section 3 describes the suggested hybrid support vector machine with grey wolf optimization (Pham et al., 2018). In Section 4, the proposed HSVMGWO and existing support vector machine, random forest, ada boost, and decision tree experimental results are compared. The proposed chapter's final remarks and future scope are found in Section 5.

2 Literature Review

This present section elaborates a review of previous studies for data classification using support vector machine-based classifiers (Emamgholizadeh et al., 2021). Galatenko et al. (2014) gave a formalised definition related to problem of choosing specification as well as building a genomic classifier for medical test systems that relies on mathematical machine learning techniques rather than biological or medical expertise (Gochhait et al., 2021). Latha (2014) presented a support vector machine (SVM) for Radial Basis Function technique with automatic analysis of Magnetic Resonance Image (MRI) (RBF). Karamizadeh Sasan et al. (2014) published a review on support vector machine (SVM), a pattern recognition and data categorization algorithm. Using the information supplied by the support vector machine (SVM), Gürbüz and Kilic (2014) built a general-purpose, rapid, and adaptive automatic disease detection system, which enhanced the success rate and reduced the decision-making time.

Chen Zhi et al. (2016) developed a support vector machine classifier (GA–SVM) based on a genetic algorithm (GA) for lymph disease diagnosis. In the first stage, GA is used to cut the 18 features in the lymph diseases dataset down to six. A support vector machine with several kernel functions, such as linear, quadratic, and Gaussian, was used as a classifier in the second stage. In the field of medical imaging, Lee et al. (2015) combined support vector machines (SVM) with Active Learning (AL) into rise prediction based on irregular classes. Jian Xiao and Sheng Hanmin (2015) proposed a Fuzzy Support Machine (FSVM) as the irregular problem class (dubbed FSVMCIP), which may be thought of as a modified FSVM with manifold regularisation for two classes, and there are two costs associated with misclassification.

QinanJia et al. (2014) proposed a method based on support vector machines to predict the exact recurrence time (SVM). In the medical domain, prognosis prediction as a regression issue is frequently used to anticipate event duration time, such as the duration time of a disease recurrence. The proposed strategy is compared to four different prognostic algorithms using the Wisconsin Breast Cancer Dataset. Jan Petrich et al. (2017) studied the effect of different CT scanners on the accuracy of high-resolution CT (HRCT) scans in categorising regional disease patterns in patients with diffuse lung disease using multicenter data. Bayesian and support vector machine (SVM) classifiers were utilised. Chang Yongjun et al. (2012) proposed a hierarchical support vector machine that distinguishes diffuse interstitial lung disease by training a binary classifier at each node in a hierarchy and allowing each classifier to use a class-specific quasi-optimal feature set to improve time and accuracy in computer-aided quantification.

Qiu Wu and Zhang proposed using discriminative information contained in a whitening transformation to convert LDA to PCA, followed by a support vector machine (SVM) technique for LDA. The kernel algorithm for the SVM solution to LDA was constructed by using the kernel technique to indirectly translate the data to kernel space.

Gudadhe et al. (2010) introduced an assistance vector machine (SVM) and artificial neural network-based decision support system for heart disease classification (ANN). To construct a decision support system for the detection of heart illness, researchers used a three- layer Multilayer Perceptron Neural Network (MLPNN). Rokach (2010) examined existing ensemble techniques and can be used as a tutorial for practitioners who want to create ensemble-based systems. The goal of ensemble approach is to combine multiple models to create a prediction model.

Based on an SVM and feature selection, Akay (2009) proposed a breast cancer screening method. In the studies, several training-test partitions of the Wisconsin Breast Cancer Dataset (WBCD) were used, which is extensively used among researchers that use machine learning algorithms for breast cancer diagnosis. To tackle the LD diagnosis problem, Wu et al. (2008) used two well-known artificial intelligence techniques: the artificial neural network (ANN) and the support vector machine (SVM). For enhancing the efficiency of support vector machines, Zhan and Shen (2005) presented a four-step training technique (SVM).

Leng Yan et al. (2016) proposed a new method for improving the speed and accuracy of SVMs with unlabelled data: one method is to build SVMs with grid points, which can be expected to speed up SVMs in the test phase; another method is to build SVMs with unlabelled data, which has been shown to improve SVM accuracy when there is very little labelled data. Kim et al. (2003) suggested utilising the SVM ensemble with bagging (bootstrap aggregation) or boosting to improve the genuine SVM's restricted classification performance. To classify heterogeneous medical data, Kumar and Arasu (2015) employed Modified Particle Swarm Optimization and Adaptive Fuzzy K-Modes (MPSO-AFKM). Mishra and colleagues. Hybrid filter-wrapper techniques for high-performance classification models were developed by (2015).

Nguyen et al. (2015a, 2015b) used wavelet transformation (WT) and interval type-2 fuzzy logic system (IT2FLS) for automated medical data classification. The IT2FLS was taught through a hybrid learning method, resulting in improved performance and reduced computational burden. Nguyen et al. (2015a, 2015b) proposed a modified Analytic Hierarchy Process (AHP)-based gene selection for microarray classification, with AHP-selected genes being used for cancer classification using the fuzzy standard additive model (FSAM). In order to handle the high-dimensional, low-sample nature of microarray data, the number of fuzzy rules is minimised using a genetic algorithm (GA).In order to reduce computational burden, Nguyen et al. (2015a, 2015b) proposed a fuzzy standard additive model (SAM) with genetic algorithm (GSAM) for healthcare data classification, high-dimensional datasets discriminative features derived by applying wavelet transformation.

Purwar and Singh (2015) tested a hybrid prediction mode with missing value imputation on three medical data sets (HPM-MI). Santhanam and Ephzibah proposed using a genetic algorithm and fuzzy logic to diagnose heart illness automatically. To increase the performance of the fuzzy inference system used to generate a classification model for the GA selected feature, the fuzzy gaussian membership function and the centroid technique were applied.

Pourpanah et al. (2019) created a fuzzy ARTMAP (FAM) and Classification And Regression Tree (CART) hybrid for medical data classification. The presented model provides consistent learning, predictions in the form of a decision tree, and the extraction of valuable explanatory rules as a decision support tool. Sindhiya and Gunasundari investigated how the genetic algorithm (GA) and other heuristic methods could be used to choose characteristics for illness identification in large dimensional datasets. Stathopoulos and Kalamboukis (2015) demonstrated how to use latent semantic analysis (LSA) to categorise large medical datasets without using the SVD solution of the feature matrix.

3 System Design

We describe work done to design and assess ways to handle missing values, attribute noise, and imbalanced class distribution in datasets to predict in this research. In this section, we'll go over a quick overview of the hybrid approach for hybrid support vector machine with grey wolf optimization (HSVMGWO) in knowledge discovery. The purpose of this stage is to select the most appropriate categorization method for a particular dataset. Because no generalisations about the best classification strategy can be made, including this phase has necessitated empirical testing of each and every prediction and analysis for a given dataset. Because the dataset under study is limited, our suggested approach uses unsupervised learning to find the best hybrid classification methods.

3.1 Support Vector Machine (SVM) Classifier:

SVM classifiers are based upon the concept of decision planes, which specify decision boundaries unambiguously. A decision plane does separate a bunch of objects with differing class memberships. Figure 3 depicts a linear classifier with a decision boundary line that distinguishes between different object types (red and blue objects separately). A certain collection of items is located on the right side of the decision plane, whereas another set of objects belonging to a different class is located on the left side. Several classification tasks are not straightforward, necessitating the use of more complicated structures in order to achieve optimal separation and accurately categorise new objects related to the test cases under consideration (Vapnik, 2000).

The data points will be categorised by allocating the system to closer of the two parallel planes that are pushed apart as far as possible using SVM. Standard SVMs, on the other hand, solve quadratic or linear programmes that take a long time to solve. As a result, PSVM classifiers are being examined, which are meant to do classification with low computational time using a simple proximity hyperplanes solution algorithm (Samanta & Nataraj, 2009). The SVM idea is explained based on the sample of a two-class dataset with p points in a q-dimensional real space, $S = \{x_i, y_i\}$, $I = 1, \ldots, p$, $x_i \in R_q$. In x_i, the q-dimensional vector

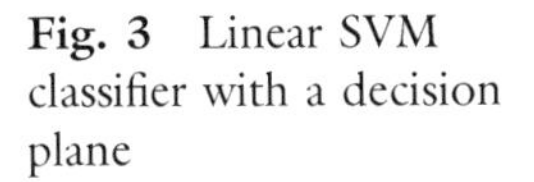

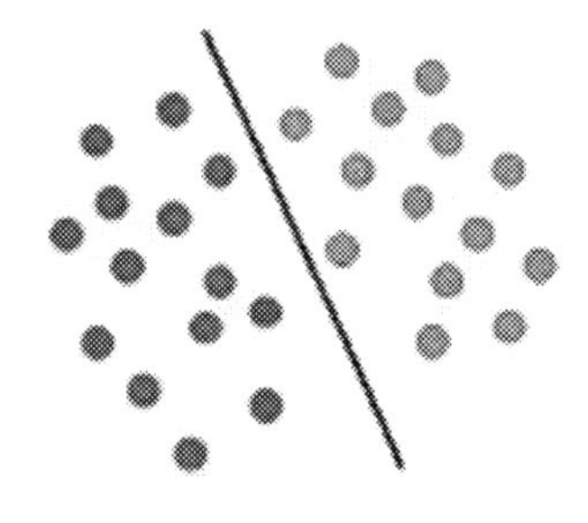

Fig. 3 Linear SVM classifier with a decision plane

category $y_i \in \{-1, 1\}$, refer to a certain class. A + and A are allocated to the equivalent datasets belonging to classes $y_i = 1$ and $y_i = -1$. The $A = [x_1, x_2, \ldots, x_p]$, T and D $=$ diag(y_i). . Variables reflect the entire dataset (y_i). The ultimate goal of SVM is to detect an optimal server hyperplane (wTx $= y$) that maximises the margin among two bounding planes (wTx $= \gamma + 1$ and wTx $= \gamma - 1$), resulting in the smallest misclassification error possible. The orientation vector w Rq is stated to be normal to the bounding planes in SVM, as is the bias, which specifies the location on the sever plane from the origin. The planes wTx $= \gamma + 1$ and wTx $= \gamma - 1$ are discovered to be included within the sets A + and A-, respectively, the margin for the border planes.

SVM is developed based on the following model:

$$\text{minimise}(w, s)\frac{1}{2}\underline{w'}\, w + Ce^T s \tag{1}$$

$$\text{subject to } D(Aw - e\gamma) + s \geq e \text{ with } s \geq 0 \tag{2}$$

To optimally separate the hyperplanes using support vectors, Eqs. (1) and (2) are considered a two optimization issue with a solution at (w, γ). The server hyperplane, which has length q and is given by, For any sample vector under consideration, it acts as a classifier.

$$x^t w - \gamma \begin{cases} > 0, \text{ then } xcA+ \\ < 0, \text{ then } xcA- \\ = 0, \text{ then } xcA + \text{or } xcA- \end{cases} \tag{3}$$

3.2 *Hybrid Support Vector Machine (SVM) with Grey Wolf Optimization (GWO)*

The data points are categorised in SVM stationed to their closeness to one of two parallel hyperplanes that distinguish datasets relationship to two various class memberships. SVM's classification performance is comparable to that of a straight forward solution technique in terms of accuracy. The suggested hybrid SVM–GWO classifiers for data classification in this research contribution combine the grey wolf optimization (GWO) proposed in this section with the linear and non-linear SVM classifiers discussed in Sect. 3.2. The proposed system design for SVM–GWO classification in each phase is depicted in Fig. 4.

For selecting the feature sets of the database under consideration, this proposed methodology integrates the concepts of SVM and GWO. The proposed self-regulated learning GWO-based SVM classifier allows users to select features from datasets to improve generalisation and conditioning of linear and non-linear classifiers. The following are the steps in the suggested method:

Step 1: Set up the SVM module's proximal planes and bias. Select the kernel functions to be used for categorization as well. Fix the search space's D dimensions.

Step 2: The output proximal hyperplanes for each member of the group are located at the SVM classifier. The training dataset is used to derive the SVM parameters 'C' and ''.

Step 3: Determine the orientation vector and separation plane location from the origin for each linear and non-linear SVM classifier.

Step 4: Now, as suggested, invoke self-regulated learning GWO.

Step 5: The fitness of individual member, in measured by the mean square error (MSE), is then assessed as follows:

$$\text{MSE} = \frac{1}{N}\sum_{I=1} E_i^2 - \frac{1}{N}\sum_{i=1}^{N}\left(y_k^i - d_k^i\right)^2. \tag{4}$$

The difference between the actual and desired outputs of the kth output neuron in the ith sample is denoted by yk and dk, where N is the number of training samples. The fitness function f is defined by the MSE in this way (x). To avoid overfitting the classifier model, each member's fitness is assessed using the mean square error (MSE) on only the validation set, rather than the whole training set.

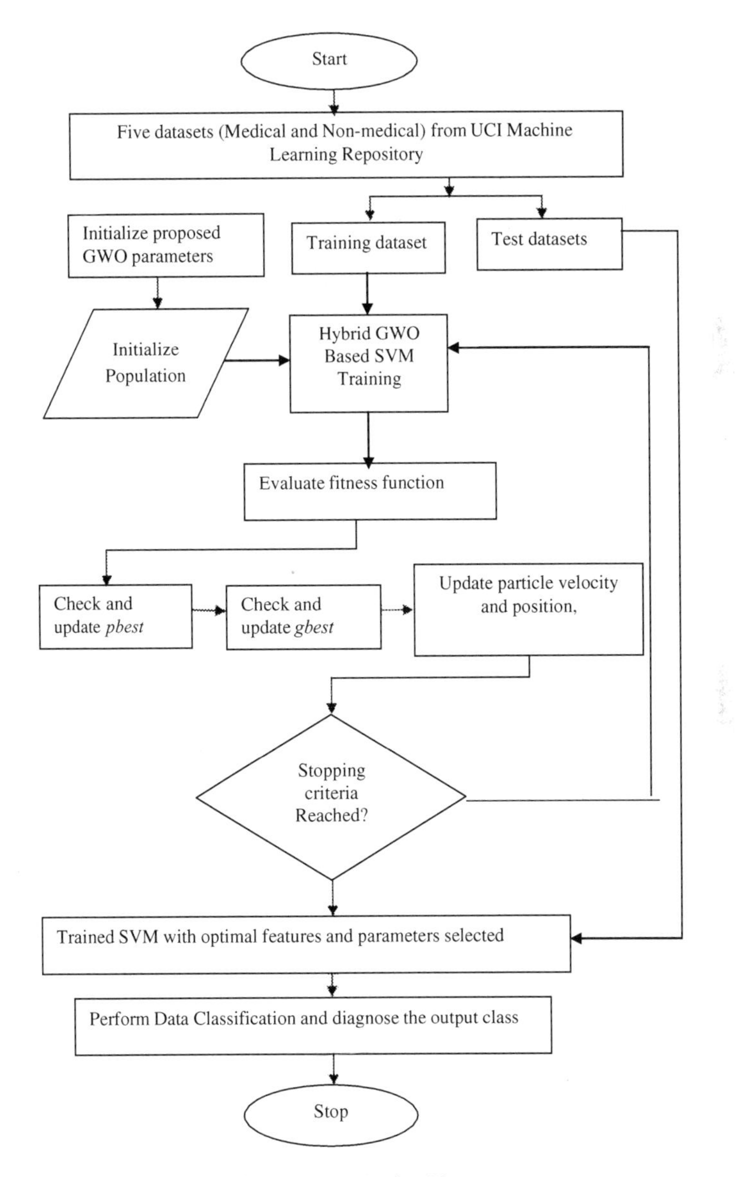

Fig. 4 Flowchart of hybrid SVM–GWO classifier

Step 6: Based on the fitness, determine the SRL acceptability.

Step 7: Update each particle's velocity and position equations.

Step 8: Stopping criteria—The algorithm continues stages 2–7 until the maximum number of iterations is reached, or until a hard threshold value is met. When the process is terminated, the values at which ideal weights with the lowest MSE are discovered are reported.

Thus, the proposed hybrid GWO with SVM classifier computes the best optimal features with the orientation vectors ('w' for linear classifier and 'v' for non-linear classifier) and the location of the separating plane from the origin " so that the fitness reaches the minimum to achieve better generalisation performance, while taking advantage of both SVM and GWO classifier advantages. The proposed hybrid SVM-based GWO classifier incorporates SVMfeatures into both linear and non-linear GWO to compute the best features with the least MSE for successful data classification.

4 Result and Discussion

PYTHON3.6IDE is used to implement the proposed methodology on an Intel(R) Core (TP) i3-2410M CPU running at 3.20GHz with 8GB RAM. This component is used to compare the selected reduct to prior data knowledge. The suggested linear and non-linear hybrid SVM–GWO classifiers' learning performance is highlighted by the performance metrics of classification accuracy, sensitivity, and specificity. From Tables 1, 2, 3 and 4, it can be deduced that the SVM–GWO outperformed other early literature classifiers like the support ada boost method, vector machine, random forest, also decision tree. Table 1 shows that the suggested SVM–GWO classifier exceeded the results of conventional techniques, indicating that it is more successful than previous classifiers. Each of the five categories of datasets has been divided towards training also testing at a 70-30 ratio. The ten-fold crosswise technique was used to validate the training dataset. The training dataset is divided into ten equal subsamples using the ten-fold crosswise procedure.

The analysis accuracy of a multivariate dataset is observed to be significantly higher than that of previous techniques. In addition, the sensitivity and specificity data that were successfully and wrongly classified are indicated to 100, indicating a flawless classification rate. Looking at the calculated simulation results for all of the other datasets in Table 1, the performance of the SVM—GWO non-linear classifier is shown to be superior than the linear SVM classifier and other existing classifiers. Because the gaussian distribution kernel in the SVM classifier explores itself in effectively detecting the separation plane and conducting classification based on the orientation vector, classification is better in the non-linear scenario.

In support vector machine, random forest, ada boost, decision tree, also proposed support vector machine (SVM) with grey wolf optimization, the accuracy comparison for shuttle-2 vs 5, yeast-0-3-5-9 vs 7-8, vowel0, cleveland-0 vs 4 and glass-0-1-4-6 vs 2 dataset in support vector machine, random forest, ada boost, decision tree, and proposed support vector machine (SVM) with grey (HSVMGWO). Figure 5 indicates that the proposed system's sensitivity outperforms all three current systems for all five categories of datasets. Shuttle-2 vs 5 dataset has a 94.7% accuracy rate, yeast-0-3-5-9 vs 7-8 has a 95.6% accuracy rate, vowel0 has a 96.3 % accuracy rate, cleveland-0 vs 4 has a 98.4 % accuracy rate, and glass-0-1-4-6 vs 2 has a 91.5 % accuracy rate. The total average accuracy was 95.3%, but the other conventional approach only got to 90.28%.

Other existing systems achieve 91.3% SVM, 88.6% random forest, 76.8% ada boost, and 95.6% decision tree on the shuttle-2 vs 5 dataset, which is lower than the proposed approach. The suggested system outperforms the yeast-0-3-5-9 vs 7-8 dataset by 87.4% of SVM, 79.4% of random forest, 81.6% of ada boost, and 78.4% of decision tree. In the vowel0 dataset, the suggested system outperforms 89.5% of SVM, 91.3% of random forest, 90.5% of ada boost, and 81.6% of decision tree. 83.6%

Table 1 SVM–GWO accuracy comparison for conventional approaches

Accuracy comparison					
Dataset	*Proposed system*	*Support vector machine*	*Random forest*	*ada boost*	*decision tree*
shuttle-2_vs_5	94.7	91.3	88.6	76.8	95.6
yeast-0-3-5-9_vs_7-8	95.6	87.4	79.4	81.6	78.4
vowel0	96.3	89.5	91.3	90.5	81.6
cleveland-0_vs_4	98.4	83.6	98.5	86.5	78.6
glass-0-1-4-6_vs_2	91.5	91.3	93.6	88.6	81.5
Average	95.3	88.62	90.28	84.8	83.14

of SVM, 98.5% of random forest, 86.5% of ada boost, and 78.6% of decision tree in the cleveland-0 vs 4 dataset are lower than the proposed system. The suggested system outperforms 91.3% of SVM, 93.6% of random forest, 88.6% of ada boost, and 81.5% of decision tree on the glass-0-1-4-6 vs 2 dataset.

In support vector machine, random forest, ada boost, decision tree, and proposed support vector machine (SVM) with grey wolf optimization, the sensitivity comparison for shuttle-2 vs 5, yeast-0-3-5-9 vs 7-8, vowel0, cleveland-0 vs 4 and glass-0-1-4-6 vs 2 dataset is explained in Table 2. (HSVMGWO). Figure 6 indicates that the proposed system's sensitivity outperforms all three current systems for all five categories of datasets. when it comes to the proposed system shuttle-2 vs 5 dataset has a sensitivity rate of 97.6%, yeast-0-3-5-9 vs 7-8 has a sensitivity rate of 95.6 %, vowel0 has a sensitivity rate of 97.3 %, cleveland-0 vs 4 has a sensitivity rate of 99.4 %, and glass-0-1-4-6 vs 2 has a sensitivity rate of 95.5 %. The total average sensitivity was 97.08 %, although other traditional approaches achieved maximum sensitivity.

Other existing systems achieve 97.6% SVM, 96.6 % random forest, 88.48 % ada boost, and 83.30 % decision tree on the shuttle-2 vs 5 dataset, which is lower than the proposed method. SVM, 89.58 % random forest, 84.15% ada boost, and 88.21% decision tree are all lower than

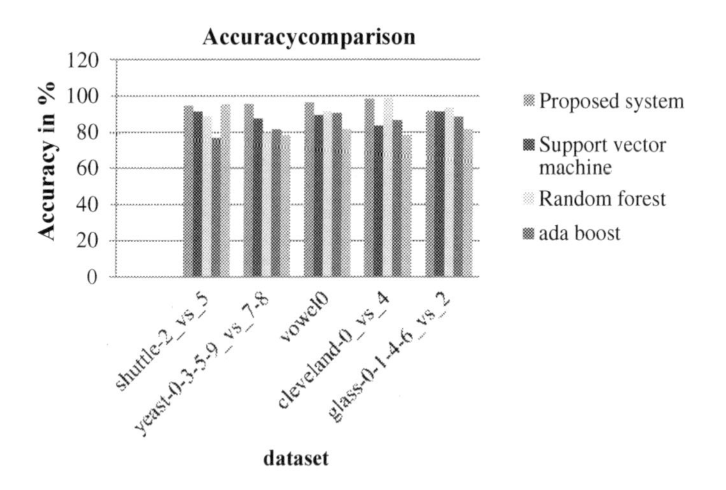

Fig. 5 SVM–GWO accuracy comparison for conventional approaches

Table 2 SVM–GWO sensitivity comparison for conventional approaches

Sensitivity comparison					
Dataset	*Proposed system*	*Support vector machine*	*Random forest*	*ada boost*	*decision tree*
shuttle-2_vs_5	97.6	96.6	88.48	79.15	83.30
yeast-0-3-5-9_vs_7-8	95.6	92.54	89.58	84.15	88.21
vowel0	97.3	87.65	84.65	86.25	89.54
cleveland-0_vs_4	99.4	94.56	93.65	88.56	93.38
glass-0-1-4-6_vs_2	95.5	93.22	96.02	84.65	91.10
Average	97.08	92.914	90.476	84.552	89.106

the proposed system in the yeast-0-3-5-9 vs 7-8 dataset. In the vowel0 dataset, the suggested system outperforms 87.65% of SVM, 84.65% of random forest, 86.25 % of ada boost, and 89.54 % of decision tree. The suggested system outperforms 94.56% of SVM, 93.65% of random forest, 88.56% of ada boost, and 93.38% of decision tree on the cleveland-0 vs 4 dataset. SVM 96.02% random forest, 84.65% ada boost, and 91.10% decision tree are all lower than the suggested system in the glass-0-1-4-6 vs 2 dataset (Fig. 6).

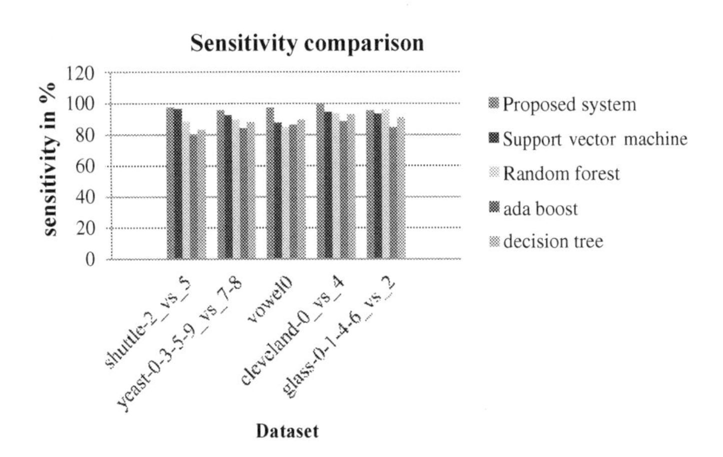

Fig. 6 SVM–GWO sensitivity comparison for conventional approaches

In support vector machine, random forest, ada boost, decision tree, and proposed support vector machine (SVM) with grey wolf optimization, the specificity comparison for shuttle-2 vs 5, yeast-0-3-5-9 vs 7-8, vowel0, cleveland-0 vs 4 and glass-0-1-4-6 vs 2 dataset in support vector machine, random forest, ada boost, decision tree, and proposed support vector machine (SVM) with grey (HSVMGWO). Figure 5 indicates that the proposed system's sensitivity outperforms all three current systems for all five categories of datasets.Shuttle-2 vs 5 dataset has a specificity rate of 91.6 %, yeast-0-3-5-9 vs 7-8 has a specificity rate of 98.6 %, vowel0 has a specificity rate of 96.3 %, cleveland-0 vs 4 has a specificity rate of 97.4 %, and glass-0-1-4-6 vs 2 has a specificity rate of 94.5 %. The proposed system achieved a total average specificity of 95.68 %, whereas other conventional approaches only achieved a maximum of 92.6%.

In comparison, other current systems achieve 87.7% SVM, 82.12% random forest,

78.84 % ada boost, and 74.4 % decision tree on the shuttle-2 vs 5 dataset, which is lower than the proposed method. The suggested system outperforms 97.2 % of SVM, 96.5 % of random forest, 93.65 % of ada boost, and 92.25% of decision tree in the yeast-0-3-5-9 vs 7-8 dataset. In the vowel0 dataset, the proposed system outperforms 93.62 % of SVM, 94.65 % of random forest, 91.91% of ada boost, and 93.35 % of decision tree. In the cleveland-0 vs 4 dataset, the suggested system outperforms 91.15 % of SVM, 89.15 % of random forest, 84.54 % of ada boost, and 86.54 % of decision tree.

Table 3 SVM–GWO specificity comparison for conventional approaches

Specificity comparison					
Dataset	*Proposed system*	*Support vector Machine*	*Random forest*	*ada boost*	*decision tree*
shuttle-2_vs_5	91.6	87.7	82.12	78.84	74.4
yeast-0-3-5-9_vs_7-8	98.6	97.2	96.5	93.65	92.25
vowel0	96.3	93.62	94.65	91.91	93.35
cleveland-0_vs_4	97.4	91.15	89.15	84.54	86.54
glass-0-1-4-6_vs_2	94.5	93.65	95.5	90.14	87.15
Average	95.68	92.664	91.584	87.816	86.738

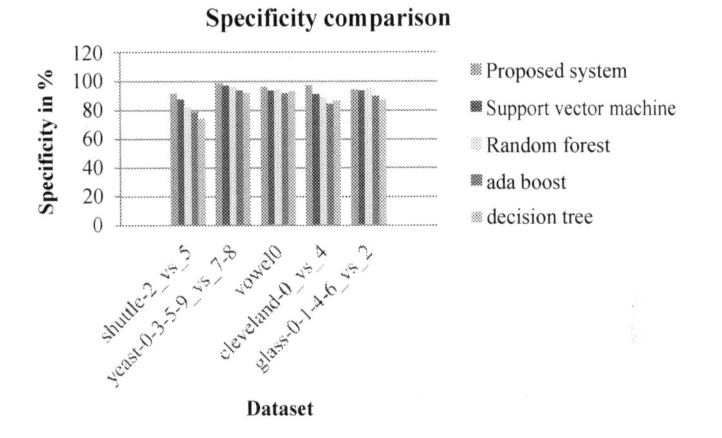

Fig. 7 SVM–GWO specificity comparison for conventional approaches

The classification time period comparison in support vector machine, random forest, ada boost, decision tree, and proposed support vector machine (SVM) with grey wolf optimization for shuttle-2 vs 5, yeast-0-3-5-9 vs 7-8, vowel0, cleveland-0 vs 4 and glass-0-1- 4-6 vs 2 dataset in support vector machine, random forest, ada boost, decision tree, and proposed support vector machine (SVM) with grey (HSVMGWO). Figure 5 indicates that the proposed system's sensitivity outperforms all three current systems for all five categories of datasets. Shuttle-2 vs 5 dataset is 31.3M.sec, yeast-0-3-5-9 vs 7-8 is 13.31M.sec, vowel0 is 8.45 M.sec, cleveland-0 vs 4 is 7.13M.sec, and glass-0-1-4-6 vs 2 is 5.74M.sec in terms of classification time duration.average total The time period for five datasets was 13.1 milliseconds, but the usual approach only achieved a minimum of 18.14 milliseconds.

In comparison, the suggested system achieves 41.14 M.sec of SVM, 46.45 M.sec of random forest, 50.51 M.sec of ada boost, and 39.14 M.sec of decision tree on the shuttle-2 vs 5 dataset. The suggested system outperforms the yeast-0-3-5-9 vs 7-8 dataset by 25.35 M.sec of SVM, 16.25 M.sec of random forest, 27.85 M.sec of ada boost, and 28.15 M.sec of decision tree. In the vowel0 dataset, SVM takes 7.95 milliseconds, random forest takes 8.21 milliseconds, ada boost takes 9.9 milliseconds, and decision tree takes 10.52 milliseconds. The suggested system is faster than 14.52 M.sec of SVM, 13.25 M.sec of random

Table 4 SVM–GWO time period comparison for conventional approaches

Time duration comparison					
Dataset	*Proposed system*	*Support vector machine*	*Random forest*	*ada boost*	*decision tree*
shuttle-2_vs_5	31.3	41.14	46.45	50.51	39.14
yeast-0-3-5-9_vs_7-8	13.31	25.35	16.25	27.85	28.15
vowel0	8.45	7.95	8.21	9.9	10.52
cleveland-0_vs_4	7.13	14.52	13.25	7.11	11.18
glass-0-1-4-6_vs_2	5.74	7.7	6.54	5.86	9.18
Average	13.186	19.332	18.14	20.246	19.634

forest, 7.11 M.sec of ada boost, and 11.18 M.sec of decision tree on the cleveland-0 vs 4 dataset. The suggested system outperforms the glass-0-1-4-6 vs 2 dataset by 7.7 M.sec of SVM, 6.54 M.sec of random forest, 5.86 M.sec of ada boost, and 9.18 M.sec of decision tree.

Tables 1, 2, 3 and 4 show the results of the hybrid support vector machine with grey wolf optimization (HSVMGWO) for classification of the selected five multivariate datasets. Reducets derived from forward feature selection for vowel0, cleveland-0 vs 4, glass- 0-1-4-6 vs 2dataset and backward feature removal approach for shuttle-2 vs 5, yeast-0-3-5-9

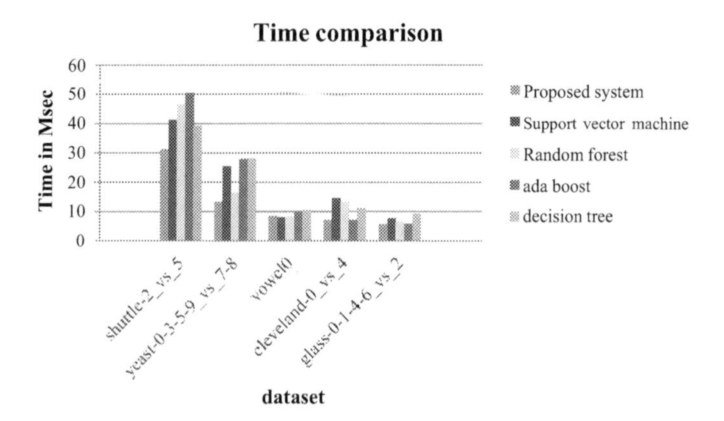

Fig. 8 SVM–GWO time period comparison for conventional approaches

vs 7-8dataset produce better performance results. The performance result indicates that the built classifier may be utilised to classify the chosen dataset in order to aid improved decision-making. Following the verification of the results, it was demonstrated that planned study can be utilised for real-time data categorization in a variety of contexts without error. The system is capable of providing real-time solutions to difficulties, and perfect achievement is achieved without delay.

5 Conclusion

This paper describes a hybrid support vector machine with grey wolf optimization (HSVMGWO) technique for classifying five different types of datasets. The suggested classifier's simulated results in this chapter show superior results, however, it should be highlighted that when used to solve a complicated data classification problem, the HSVMGWO is easily stuck in local optimization during the search process for effective features for classification. This can be seen in the datasets shuttle-2 vs 5, yeast-0-3-5-9 vs 7-8, vowel0, cleveland-0 vs 4, and glass-0-1-4-6 vs 2, where the suggested algorithm got stuck with local minima numerous times and had to be elevated to give classification results. In the future, knowledge engineers may be able to create efficient decision support systems in real-world scenarios employing hybrid classification methodologies including two or more classifiers. Hybrid optimization techniques and bio-inspired artificial intelligence approaches will generate stronger classifier models in the future, which can be employed in the design and development of decision support systems to increase efficiency.

References

Akay, M. F. (2009). Support vector machines combined with feature selection for breast cancer diagnosis. *Expert Systems with Applications, 36*(2), 3240–3247.

Beckett, C., Eriksson, L., Johansson, E., & Wikström, C. (2017). Multivariate data analysis (MVDA). *Pharmaceutical quality by design: A practical approach*, 201–225.

Chang, Y., Kim, N., Lee, Y., Lim, J., Seo, J. B., & Lee, Y. K. (2012). Fast and efficient lung disease classification using hierarchical one-against-all support vector machine and cost-sensitive feature selection. *Computers in Biology and Medicine, 42*(12), 1157–1164.

Chen, Z., Lin, T., Tang, N., & Xia, X. (2016). A parallel genetic algorithm based feature selection and parameter optimization for support vector machine. *Scientific programming, 2016*.

Emamgholizadeh, S., & Mohammadi, B. (2021). New hybrid nature-based algorithm to integration support vector machine for prediction of soil cation exchange capacity. *Soft computing, 25*, 13451–13464. https://doi.org/10.1007/s00500-021-06095-4

Galatenko, V. V., Lebedev, A. E., Nechaev, I. N., Shkurnikov, M. Y., Tonevitskii, E. A., & Podolskii, V. E. (2014). On the construction of medical test systems using greedy algorithm and support vector machine. *Bulletin of Experimental Biology and Medicine, 156*(5), 706–709.

Gochhait, S. et al. (2021). Data interpretation and visualization of COVID-19 cases using R programming. *Informatics in Medicine Unlocked*, *26*(6), Elsevier, ISSN: 0146–4116.

Gudadhe, M., Wankhade, K., & Dongre, S. (2010, September). Decision support system for heart disease based on support vector machine and artificial neural network. In *2010 International Conference on Computer and Communication Technology (ICCCT)* (pp. 741–745). IEEE.

Gürbüz, E., & Kılıç, E. (2014). A new adaptive support vector machine for diagnosis of diseases. *Expert Systems, 31*(5), 389–397.

Kalimuthu, S. (2021). Sentiment analysis on social media for emotional prediction during COVID-19 pandemic using efficient machine learning approach. *Computational intelligence and healthcare informatics*, 215.

Kalimuthu, S., Naït-Abdesselam, F., & Jaishankar, B. (2021). Multimedia data protection using hybridized crystal payload algorithm with chicken swarm optimization. In *Multidisciplinary approach to modern digital steganography* (pp. 235–257). IGI Global.

Karamizadeh, S., Abdullah, S. M., Halimi, M., Shayan, J., & javadRajabi, M. (2014, September). Advantage and drawback of support vector machine functionality. In *2014 international Conference on Computer, Communications, and Control Technology (I4CT)* (pp. 63–65). IEEE.

Kim, H. C., Pang, S., Je, H. M., Kim, D., & Bang, S. Y. (2003). Constructing support vector machine ensemble. *Pattern Recognition, 36*(12), 2757–2767.

Kumar, R. S., & Arasu, G. T. (2015). Modified particle swarm optimization based adaptive fuzzy k-modes clustering for heterogeneous medical databases.

Latha, P. (2014). SVM based automatic medical decision support system for medical image. *Journal of Theoretical & Applied Information Technology, 66*(3).

Lee, S. H., Bang, M., Jung, K. H., & Yi, K. (2015, June). An efficient selection of HOG feature for SVM classification of vehicle. In *2015 International Symposium on Consumer Electronics (ISCE)* (pp. 1–2). IEEE.

Leng, Y., Sun, C., Xu, X., Yuan, Q., Xing, S., Wan, H., & ...& Li, D. (2016). Employing unlabeled data to improve the classification performance of SVM, and its application in audio event classification. *Knowledge-based systems, 98*, 117–129.

Liu, R., Peng, J., Leng, Y., Lee, S., Panahi, M., Chen, W., Zhao, X. (2021) Hybrids of support vector regression with grey wolf optimizer and firefly algorithm for spatial prediction of landslide susceptibility. *Remote Sensing*, *13*(24), 4966. https://doi.org/ https://doi.org/10.3390/rs13244966

Nguyen, T., Khosravi, A., Creighton, D., & Nahavandi, S. (2015a). Classification of healthcare data using genetic fuzzy logic system and wavelets. *Expert Systems with Applications, 42*(4), 2184–2197.

Nguyen, T., Khosravi, A., Creighton, D., & Nahavandi, S. (2015b). Medical data classification using interval type-2 fuzzy logic system and wavelets. *Applied soft computing, 30*, 812–822.

Orriols-Puig, A., & Bernadó-Mansilla, E. (2009). Evolutionary rule-based systems for imbalanced data sets. *Soft Computing, 13*(3), 213–225.

Petrich, J., Gobert, C., Phoha, S., Nassar, A. R., & Reutzel, E. W. (2017, August). Machine learning for defect detection for PBFAM using high resolution layerwise imaging coupled with post-build CT scans. In *Proceedings of the 27th international Solid Freeform Fabrication Symposium.*

Pham, B. T., Tien Bui, D., & Prakash, I. (2018). Bagging based support vector machines for spatial prediction of landslides. *Environment and earth science, 77*, 146.

Pourpanah, F., Lim, C. P., & Hao, Q. (2019). A reinforced fuzzy ARTMAP model for data classification. *International Journal of Machine Learning and Cybernetics, 10*(7), 1643–1655.

Purwar, A., & Singh, S. K. (2015). Hybrid prediction model with missing value imputation for medical data. *Expert Systems with Applications, 42*(13), 5621–5631.

Qinan, J., Lei, M., Jianfeng, H., QingQing, Y., & Jun, Z. (2014). A primary study for cancer prognosis based on classification and regression using support vector machine. In *Frontier and future development of information technology in medicine and education* (pp. 909–920). Springer, Dordrecht.

Qiu, J., Wu, Q., Ding, G., Xu, Y., & Feng, S. (2016). A survey of machine learning for big data processing. *EURASIP Journal on Advances in Signal Processing, 2016*(1), 1–16.

Rokach, L. (2010). *Ensemble-Based Classifiers. Artificial Intelligence Review, 33*(1), 1–39.

Ros, G., Sellart, L., Materzynska, J., Vazquez, D., & Lopez, A. M. (2016). The synthia dataset: A large collection of synthetic images for semantic segmentation of urban scenes. In *Proceedings of the IEEE conference on computer vision and pattern recognition* (pp. 3234–3243).

Samanta, B., & Nataraj, C. (2009). Use of particle swarm optimization for machinery fault detection. *Engineering Applications of Artificial Intelligence, 22*(2), 308–316.

Sheng, H., & Xiao, J. (2015). Electric vehicle state of charge estimation: Nonlinear correlation and fuzzy support vector machine. *Journal of Power Sources, 281*, 131–137.

Sindhiya, S., & Gunasundari, S. (2014, February). A survey on genetic algorithm based feature selection for disease diagnosis system. In *Proceedings of IEEE International Conference on Computer Communication and Systems ICCCS14* (pp. 164–169). IEEE.

Stathopoulos, S., & Kalamboukis, T. (2015). Applying latent semantic analysis to large-scale medical image databases. *Computerized medical imaging and graphics, 39*, 27–34.

Vapnik, V., & Chapelle, O. (2000). Bounds on error expectation for support vector machines. *Neural Computation, 12*(9), 2013–2036.

Wu, C. H., Chou, H. J., & Su, W. H. (2008). Direct transformation of coordinates for GPS positioning using the techniques of genetic programming and symbolic regression. *Engineering Applications of Artificial Intelligence, 21*(8), 1347–1359.

Zhan, Y., & Shen, D. (2005). Design efficient support vector machine for fast classification. *Pattern Recognition, 38*(1), 157–161.

CHAPTER 8

Bioinformatics and Its Application in Computing Biological Data

Sonali Patil and Annika Durve Gupta

1 Introduction

Bioinformatics is a study of interdisciplinary that creates software and methods tools for analysing biological data, specifically complex and large data sets. It is a study that combines biology, information engineering, computer science, mathematics, statistics, and even reconstruction, pattern recognition, simulation, machine learning, iterative approaches, and molecular algorithms or modelling to analyse and interpret biological data. For statistical and mathematical in silico analyses of biological questions, bioinformatics has been used. One use of bioinformatics is the examination of molecular sequences and genomic data. The goal of bioinformatics, which is a combination of life sciences branches,

S. Patil (✉)
Department of Bioanalytical Sciences, B. K. Birla College, Kalyan, Maharashtra, India
e-mail: sonali1386@gmail.com

A. D. Gupta
Department of Biotechnology, B. K. Birla College, Kalyan, Maharashtra, India

S. Dutta and S. Gochhait (eds.), *Information Retrieval in Bioinformatics*,
https://doi.org/10.1007/978-981-19-6506-7_8

is to develop methodologies and tools for organising, storing, systematising, visualising, annotating, querying, understanding, and interpreting large amounts of biological data, as well as managing large amounts of data.

2 Goals

Bioinformatics encompasses both biological research that incorporates computer programming and a collection of regularly used analysis "pipelines", specifically in genomics field. Bioinformatics is commonly used to identify candidate genes and single nucleotide polymorphisms (SNPs). Such identification is usually done in order to better understand unique adaptations, genetic basis of disease, attractive characteristics (in agricultural fields), or population variances. Bioinformatics, often known as proteomics, is an informal term for the study of the organisational principles inside nucleic acid and protein sequences.

Gene discovery, drug design, sequence alignment, genome assembly, gene expression, protein structure alignment, drug discovery, protein structure prediction, and prediction of protein–protein interactions, evolution modelling, genome-wide association studies, and cell division/mitosis are just a few of the major research efforts in the field (Frantzi et al., 2019).

Featured sub-disciplines within computational biology and bioinformatics include:

- Implementing and creating computer programmes which allows effective access to, utilisation and management various types of data
- Developing new algorithms (mathematical formulas) and statistical measures for assessing relationships between members of large data sets. Methods to find a gene within a sequence and/or predict protein structure, as well as cluster protein sequences into families of related sequences, are all available.
- Current research in molecular research and genomics techniques, as well as advances in information technology, have combined to create a large volume of molecular biology data over the previous few decades. The word "bioinformatics" refers to the application of mathematical and computer approaches to better understand biological processes.

Mapping and aligning DNA/protein sequences, analysing DNA and protein sequences to compare them, and producing and displaying 3D models of protein structures are all bioinformatics techniques (Benson et al., 1998).

3 Relationship to Other Disciplines

Biological computation and bioinformatics are related but distinct disciplines of science. Sometimes, there is confusion between computational biology and biology. Biological computation uses biology and bioengineering to create biological computers, whereas bioinformatics uses computation to learn more about biology. Computational biology and bioinformatics are two fields that study the sequences of RNA, DNA, and proteins in biological data. A major part of the Human Genome Project and substantial advances in DNA sequencing technology resulted in the rapid growth of the field of bioinformatics in the mid-1990s (Allaby & Woodwark, 2004, Porter & Hajibabaei, 2021). Analysing biological data and producing meaningful information that is used by computer algorithms based on artificial intelligence, graph theory, data mining, soft computing, computer simulation, and image processing. It is based on theoretical foundations such as discrete system theory, mathematics, control theory, statistics, and information theory (Cantor, 1998).

4 Databases

Bioinformatics provides different databases and tools for analysing biological data.

4.1 *The Use of Software and Tools*

Bioinformatics software tools range from command-line utilities, i.e. simple to even complex graphical applications also standalone web services, all of which are offered from bioinformatics companies or government agencies.

4.1.1 *Bioinformatics Software that is Free and Open Source*

Since the 1980s, a number of free and open-source software applications have gained popularity. All research groups have been able to contribute to both bioinformatics and the wide range of open-source software available like Bioconductor, BioPerl, Biopython, and many others by combining new algorithms for determining the potential for innovative in silico experiments, emerging types of biological readouts, and freely available open code bases. Community-supported plug-ins and idea incubators are often used in commercial applications. Besides helping with de facto standards and shared object models, they may also be able to assist with the integration of biodata, which could be a big help in the future.

In order to facilitate the use of algorithms, data, and computing resources on machines in different parts of the world, SOAP and REST-based interfaces have been developed for many bioinformatics applications. There is a significant benefit in that end users are relieved of the burden of maintaining software and databases.

The EBI (Biological Sequence Analysis) defines three types of basic bioinformatics services: SSS (Sequence Search Services), MSA (Multiple Sequence Alignment), and BSA (Fundamental Sequence Alignment). It is possible to build standalone, web-based, or integrated bioinformatics workflow management systems based on the use of standalone tools with shared data formats.

4.1.2 Bioinformatics Workflow Management Systems

An application for managing bioinformatics workflows, commonly known as workflows, refers to a collection of computational and data manipulation procedures known as a workflow. It is intended to provide an easy-to-use environment that allows individual application scientists to easily create their own workflows within these systems;

- Make the process of sharing and reusing workflows among scientists easier.
- Allow scientists to monitor the provenance of their workflows by providing interactive tools enabling scientists to run workflows and view outcomes in real-time.

Galaxy, UGENE, Kepler, Taverna, HIVE, and Anduril are some of the platforms that provide this service.

5 Applications of Bioinformatics

Bioinformatics is being used in fields like microbial genome applications, medicines, agriculture, and veterinary sciences.

5.1 Microbial Genome Applications

All genetic material from bacteria and eukaryotes, including chromosomal and extrachromosomal components, is contained in microbial genomes. This is a crucial field when it comes to Bioinformatics applications. Bioinformatics tools can help with DNA sequencing for usage in domains including health and energy, in addition to analysing genome assembly (Mann et al., 2021).

5.1.1 Waste Cleanup

For application in a range of sectors, millions of dangerous compounds have been developed. These compounds are frequently discharged into the environment as a result of human activities, contaminating water and soil. Furthermore, many chemicals persist in the environment, causing major health concerns to living organisms; it is vital that these toxins be removed from the environment. The breakdown of xenobiotic or chemicals substances by plants and bacteria is known as biodegradation (Sadraeian & Molaee, 2009). Toxic compounds are degraded by biodegrading bacteria through co-metabolism or mineralisation. Microbes totally breakdown harmful substances by using them as carbon and energy sources during the mineralisation phase, whereas poisonous compounds are biotransformed into less toxic ones during co-metabolism. Microbial remediation is an unique method of removing toxic substances from the environment (Arora et al., 2009). Many microorganisms have been isolated that can use harmful chemicals as their primary source of energy and carbon, with many of them using microbial enzymes like monooxygenases, dioxygenases, reductases, deaminases, and dehalogenases to break complicated chemical compounds down to carbon dioxide and water. To improve bioremediation efficiency, the genes encoding these enzymes have been found in a variety of microorganisms and cloned into bacteria. The decomposition of a certain harmful chemical necessitates the use of a specific microbe, which is dependent on the chemical's structure and the availability of enzyme systems in bacteria. As a result, understanding chemicals (classification, distribution, identification, toxicity, environmental features, and related dangers) as well as their microbial biodegradation (xenobiotics degrading bacteria genes, proteins, and enzymes) can help in bioremediation. Bioinformatics has been integrated into all fields of biological sciences, providing a platform for researchers to construct valuable computational tools for human and environmental well-being. Bioinformatics has been merged with biodegradation throughout the last few decades, and several bioinformatics tools that are useful in the subject of biodegradation have been created. Databases, prediction of biodegradation pathway, prediction systems for chemical toxicity, and next-generation sequencing are examples of these technologies (Greene, 2002). Biodegradative databases collect information about chemical biodegradation, such as xenobiotic-degrading microorganisms, toxic chemical metabolic breakdown pathways, enzymes, and genes involved in biodegradation. Biodegradation Network-Molecular Biology Database (Bionemo), MetaCyc, The University of Minnesota Biocatalysis/Biodegradation Database (UM-BBD), an OxDBase Database of biodegradative oxygenases, and BioCyc are among these databases (Caspi et al., 2012).

5.1.2 *Biotechnology*

Global economic and social challenges are being addressed by advances in molecular modelling, pharmaceutical discovery, disease characterisation, forensics, clinical health care, and agriculture in the biotechnology industry. Bioinformatics has reached unprecedented heights among the biological disciplines as a result of public trust in biotechnology and biotechnology's advancement. Automatic gene identification, genome sequencing, prediction of gene function, phylogeny, drug design and development, protein structure prediction, vaccine development, organism identification, comprehending genomic and gene complexity, protein functionality, structure, and folding and other bioinformatics applications exist to speed up research in the field of biotechnology. The use of bioinformatics in research allows researchers to complete long-term research projects quickly, such as genome mapping. The future demands of biotechnology will also be met by bioinformatics innovation. The role of bioinformatics in many biotechnology disciplines has been discussed, including genomics, drug design, proteomics, and environmental biotechnology.

Genomics

Genomics refers to the study of genes and their expression. This subject generates a large amount of data regarding gene sequences, their interrelationships, and their functions. Bioinformatics plays a crucial role in managing this massive amount of data. It is becoming easier and easier to detect systemic functional behaviour, as more complete genome sequences for more animals become available, through bioinformatics. Thompson et al. (1994) assert that bioinformatics is critical in structural genomics, nutritional genomics, and functional genomics.

Proteomics

The study of the function, structure, and interactions of proteins produced by a tissue, cell, or organism is known as proteomics. It includes methods in biochemistry, genetics, and molecular biology. Massive volumes of data on protein profiles, protein activity patterns, protein–protein interactions, and organelle compositions have been generated using advanced biological techniques. This huge amount of data can be managed and accessed using bioinformatics software, databases, and tools. Image analysis of 2D gels, peptide mass fingerprinting, and fingerprinting of peptide fragmentation are just a few of the techniques established in the field of proteomics so far (Hanash, 2003).

Comparative Genomics

In comparative genomics, bioinformatics is used to determine the genetic structural and functional relationships between various biological species.

Transcriptomics

Transcriptomics is the study of groupings of all messenger RNA molecules in a cell (Marini et al., 2021). This is also known as Expression Profiling, and it comprises utilising a DNA microarray to measure the level of mRNA expression in a specific cell group. Microarray technology creates thousands of data values in a single run, while a single experiment necessitates hundreds of runs. To analyse such vast amounts of data, a variety of software packages are used. For transcriptome analysis, bioinformatics is used in this way, allow to determine mRNA expression levels. RNA-sequencing (RNA-seq) has also been included in the transcriptomics category (Eagles et al., 2021). The quantity and existence of RNA in a sample at a certain moment are determined using next-generation sequencing. It's used to investigate how the cellular transcriptome evolves throughout time.

Cheminformatics

Cheminformatics (chemical informatics) is the study of chemical substance information storage, indexing, searching, retrieval, and application. It comprises the logical organisation of chemical data in order to make chemical structures, characteristics, and interactions more accessible. It is theoretically possible to design a chemical with the required properties, detect and structurally modify a natural product, and test its therapeutic effectiveness using computer algorithms employing bioinformatics. Cheminformatics analysis includes procedures like as grouping, similarity searches, QSAR modelling, virtual screening, and others.

Gene Expression

Gene expression regulation allows researchers to use genetic data to construct molecular technologies that is the basis of functional genomics, and it that can count the number of genes that are currently being transcribed in each cell at any given time (e.g. gene expression arrays).

5.1.3 Climate Change Studies

Because of the accelerated sea level rise, loss of sea ice, and more and longer powerful heat waves, climate change is a global problem. Bioinformatics may be able to help solve this problem by sequencing microbial genomes, which reduces carbon dioxide as well as other greenhouse gas levels. This contributes significantly to the stabilisation of global climate change. In the bioinformatics sector, more location-specific research is needed, taking into account the microbes of particular region and their ability to reduce CO_2 (Sinha, 2015).

5.1.4 Bio-Weapon Creation

One of the most serious dangers to home and land security will continue to be the deliberate deployment of traditional or combinatorial bioweapons. In the domains of synthetic, molecular, and computational biology, the misuse of dual-use and how-to procedures and techniques could lower the technological hurdles to launching assaults, even for small organisations or individuals. A variety of biodefense techniques are being developed using bioinformatics. On the other hand, existing algorithms have failed to transform pathogen genetic data into standardised diagnoses, broad-spectrum medicines, or rational vaccine development.

Bioinformatics has a limited impact, despite its potential. More than a dozen biodefense databases and information exchange platforms lack interoperability and a common layer, restricting biodefense enterprises' scalability and development. As a result, the development of computational biology applications must be prioritised in order to leverage next-generation genome sequencing for forensic operations, medical intelligence, mitigation, and biothreat awareness.

According to the European Bioinformatics Institute, scientists had sequenced the genomes of 3139 viruses, 1016 plasmids, and 2167 bacteria as of December 2012, some of which are publicly available on the internet. Scientists will soon be able to produce infections by making synthetic viruses, synthetic genes, and possibly entirely new creatures, thanks to the availability of full genomes and the afore-mentioned breakthroughs in gene synthesis.

Furthermore, the exponential growth in computer capacity, along with public access to genetic knowledge and biological equipment, as well as a lack of government oversight, raises concerns about biowarfare from non-military sources. According to the US government, terrorist networks are recruiting scientists capable of producing bioweapons, which "has prompted countries to be more transparent about their attempts to clamp down on the threat of bioweapons".

Alternative Energy Sources

One of the greatest difficulties of the twenty-first century is finding a renewable and affordable source of energy. Plants have long been employed as a renewable energy source, and they remain one of the most promising prospects in this field. Plant biomass growth and quality must be constantly enhanced in order for plants to continue to provide a cost-effective and renewable source of energy. Recent advancements in genomics, aided by the advent of high-throughput sequencing and genotyping tools, have opened up new options for plant breeding and biotechnology development. The continual development of computers and bioinformatics tools is required to analyse the large amounts of data generated by these current genomics platforms. To employ genomics for finding alternative energy sources, bioinformatics methods for gene expression analysis with RNA-seq and for SNP genotyping can be applied.

It will make it possible. Engineering strategies are anticipated by the availability of large sequenced genomes, metabolic pathway reconstruction, functional genomics investigations, the development of in silico models at the genome scale, and synthetic biology approaches.

Renewable energy sources have a lot of potential in terms of biofuels. Bioinformatics is required to understand and analyse biofuel manufacturing processes. Genetically engineered microalgal strains with suitable lipid content have been made possible with recent advances in algal genomics and other "omics" approaches.

Bioinformatics can also be used to discover alternative energy sources in bacterial organisms. Bioinformatics is being used to examine the genome of *Chlorobium tepidum*, a bacteria with an extremely large genome.

5.2 Medicines

In medicine, bioinformatics has several uses, including gene research, medicinal development, and prevention. Medical applications of bioinformatics are:

- **Pharmaceuticals:** Pharmaceutics research has relied heavily on bioinformatics researchers, particularly in the field of infectious diseases. Personalised medical research has also advanced thanks to bioinformatics, with new therapies based on a person's genetic profile being discovered.
- **Prevention:** By analysing disease patterns, community healthcare infrastructure, illness causes, and so on, bioinformatics, like pharmaceuticals, can be integrated into preventative medicine to produce preventative medicine.
- **Therapy:** Bioinformatics can be effective in gene therapy, particularly for single genes that have been harmed. Genetics scientists have studied this application of bioinformatics and discovered that using bioinformatics, one's genetic profile can be improved.

5.2.1 Molecular Medicine

In modern biology, bioinformatics is the study of two fundamental information flows. The first is the transmission of genetic information from an individual organism's DNA to a population with qualities that are similar to those of the same species. The second is the flow of experimental data from observed biological phenomena to explanation models, which is then followed by more tests to put the models to the test. The organisation of DNA sequence and protein 3D structural data collections was one of the first initiatives in bioinformatics in the 1960s and 1970s. With the development of biological investigations that produce vast volumes of data quickly, it has grown into a thriving academic and corporate sector (such as the multiple genome sequencing projects, the large-scale analysis of gene expression, and the large-scale analysis of protein–protein interactions). Clinical medicine (including clinical medical information systems) has long been affected by basic biological science, and a new generation of epidemiologic, prognostic, diagnostic, and therapeutic tools is emerging. Over the next decade, bioinformatics activities that appear to be solely focused on basic research are expected to become increasingly relevant in clinical informatics. DNA sequence information and annotations, for example, will become more widespread in medical records. Clinical information systems will soon incorporate bioinformatics technologies established for research. The focus of genetic disorder research is turning away from single genes and towards uncovering networks of genes at cellular level, unravelling their intricate connections, and establishing their role in disease. As a result, a new era of individually personalised treatment will emerge. Bioinformatics will aid and guide clinical researchers and molecular biologists in taking advantage of the advantages of computational biology. Clinical research teams who can seamlessly move from clinical practice to the laboratory bench to the use of these powerful computational tools will be the most prolific in the coming decades.

Personalised Medicine

The medicine is a sort of treatment that is personalised to each individual's genetic composition. Personalised medicine is a type of medical care in which each patient's treatment is individually adjusted to meet their specific needs. It is conceivable since we are genetically diverse from one another. There are two important keys in the concept. To begin, medical research attempts highlight how personalised medicine is. Shifting medicines focus from reaction to prevention, selecting the optimal therapy, reducing the length and cost of clinical trials, lowering the overall estimated cost of health care, and lowering the likelihood of adverse drug reactions are all attempts (Zhang and Hong, 2015).

The second aspect is information technology's rapid improvement, which has led to the development of novel technologies for decoding human genomes, large-scale genetic variation research, and medical informatics. Due to rapid development of new sequencing technologies, sequencing has moved to a new level. Various sequencing techniques have arisen since 2003, when human genome project was completed. As a result, whole genome sequencing is now less expensive, reducing from $2.3 billion in 2003 to just $1000 in 2016, and this trend may continue in the next years or decades. Artificial intelligence (AI) and the development of more powerful algorithms to produce learning machines play a role in efficiently managing data sets and accurately projecting outcomes. AI has already aided in the development of medical technology, from medication development to the development of assisted systems for clinics, as a tool that can replicate brain-based cognitive function. Breast cancer diagnosis is one important application of deep learning based on this concept.

Personalised medicine uses bioinformatics to analyse data from genome sequencing or microarray gene expression studies to detect mutations that may affect treatment response or prognosis. The Randomized Algorithm and CADD can be useful tools (Cello et al., 2002).

Knowledge-based information is being incorporated into new algorithms. SIFT is an evolutionary knowledge-based approach for predicting mSNPs. SIFT uses a multiple sequence alignment between homolog proteins to score the normalised probabilities for all possible substitutions, whereas PolyPhen uses different sequence-based features and a position-specific independent counts (PSICs) matrix from multiple sequence alignment to predict the impact of mSNPs (Chenna et al., 2003). The PANTHER algorithm uses a library of protein family to predict hazardous mutations. 3D structural features can predict disease-related mSNPs. Knowledge-based information has increased the predictability of algorithms by more than 80%. For example, SNPs & GO is a functional information-based technique that uses log-odd scores derived from annotation in Gene Ontology (GO) words as input. MutPred assesses the likelihood of structure and function gain or loss as a result of mutations and identifies their impact.

5.2.2 *Preventative Medicine*

All physicians practise preventive medicine in order to keep their patients healthy. It's also a one-of-a-kind medical field recognised by the American

Board of Medical Specialties (ABMS). Preventive medicine focuses on individuals, communities, and defined groups. To understand the patterns and causes of diseases and health, it uses a variety of research methods, such as bioinformatics, biostatistics, and epidemiology. As well as treating obesity and blindness, it can also help with weight loss.

To gain a better understanding of the population's health and disease patterns and causes, as well as to integrate this information into disease-prevention methods, research has been done. Multi-site, longitudinal cohort studies are part of the research, and the faculty monitors a number of investigator-initiated studies. Preventive medicine, often known as preventive care, is a set of practices aimed at preventing diseases rather than curing and treating their symptoms. Curative and palliative medicine, as well as applied public health measures, can all be utilised to achieve this goal.

The screening of neonates immediately after delivery for health issues, such as metabolic disorders or genetic diseases, that are detectable but not visible clinically in the newborn period, is an example of preventive medicine.

Bioinformatics enables the collection and processing of data, as well as the standardisation and harmonisation of data for scientific discovery and the fusion of different data sources. Interoperability (the establishment of an informatics system that allows access to and use of data from many systems) will make scientific discoveries and vocations easier, as well as potential for public health interventions. The National Cancer Institute (NCI) has interoperable Cancer Biomedical Informatics Grid (caBIG) which is the example of technologies used by population scientists. Progress necessitates more than just tools. There are still issues, such as a lack of common data standards, private data access hurdles, and challenges pooling data from different studies. To overcome these obstacles, population scientists and informaticists are proposing new and creative solutions.

To develop screening tests to diagnose diseases at an early stage, researchers employ bioinformatics methods to analyse genomes, proteomics, and metabolomics data.

The most recent example of bioinformatics-based preventative trials is COVID-19 (Gochhait et al., 2021). The discovery of a large number of coronaviruses, as well as coronavirus genomes that have been sequenced, has proven to be an unusual occurrence for doing bioinformatics and genomics studies on the virus family. The coronavirus is with large genome when compared to other RNA viruses (26.4 kb to 31.7 kb). Coronavirus has a high G + C content, ranging between 32 and 43 per cent. Based on their phylogenetic links, coronaviruses are divided into three lineages: Alpha coronavirus, Beta coronavirus, and Gamma coronavirus. Coronaviruses are classified into three lineages based on their phylogenetic relationships: Alpha coronavirus, Beta coronavirus, and Gamma coronavirus. As a result, the Beta coronavirus has been separated into four subdivisions: A, C, B, and D.

Coronaviruses are well-known infections that can affect both animals and humans. Between conserved genes and in the nucleocapsid gene's posterior, a varied number of small ORFs can be discovered. According to these uses, bioinformatics studies have been involved in the prevention of COVID-19. A large number of bioinformatics NGS experiments are being conducted for COVID-19 research. Using bioinformatics techniques and software, we can predict protein structure and determine which genes are relevant to coronavirus.

5.2.3 *Gene Therapy*

The field of gene therapy involves implanting genetic components into diseased cells to treat, cure, and prevent disease. The use of Bioinformatics in Gene Therapy includes identifying cancer types, analysing protein targets, and assessing microRNA. The procedure of replacing a patient's damaged genes having functional one in their cells is known as gene therapy. Gene therapy isn't widely used since creating a generic gene therapy strategy is challenging due to the fact that everyone's genetic makeup is different. Based on their DNA sequence, bioinformatics will help determine the right gene target location for everyone.

5.2.4 *Drug Development*

Drug discovery is the most important application of bioinformatics. Computational biology, a subset of bioinformatics, aids researchers in deciphering disease causes and validating innovative, cost-effective treatments. Infectious diseases are the greatest cause of death in children and adolescents around the world. According to the World Health Organization, over 13 million people are being killed because of infectious diseases each year. Molecular modelling and simulation can help speed up the process of identifying therapeutic targets and screening drug candidates, and more effective and safer drugs can be generated. In the case of the COVID-19 epidemic, bioinformatics can be used to develop a low-cost, effective medication (Imming et al., 2006, Sharma et al., 2021). The Swiss Institute of Bioinformatics maintains http://click2drug.org/, which contains a comprehensive collection of tools, web services, and databases, directly connected to drug discovery. These are roughly divided into Thirteen categories: (1) databases, (2) molecular modelling and simulation, (3) chemical structure representations, (4) the structure of a protein is inferred via homology modelling, which is guided by a homologue of known structure, and (5) Docking, (6) Prediction of binding sites, (7) The structure of a protein can be inferred via homology modelling, which uses a homologue of known structure as a guide, (8) Drug Candidate Screening, (9) Prediction of Drug Targets, (10) Free Energy Binding Estimation, (11) Ligand Design, (12) QSAR, and (13) ADME Toxicity (Anderson, 2003). Many powerful and free software products are funded by well-known institutions. These include databases like ChEMBL and Swiss Sidechain, software tools like UCSF Chimera, which is not only a 3D visualisation tool but also a platform for structural biology software developers, Swiss Bioisostere for ligand design, Swiss Similarity for virtual screening, Swiss SideChain, Swiss Target Prediction to facilitate experiments that expand the protein repertoire by introducing new proteins, and Swiss Similarity for virtual screening, Swiss Bioisostere for PyMOL and CHARMM (Schrödinger) are examples of commercial tools that often include free versions for students and teachers (Stoesser et al., 1998).

5.2.5 *Antibiotic Resistance*

Antibiotic drug resistance is a major worry all over the world. Antibiotics put selective pressure on the propagation of resistance genes by allowing bacterial isolates to exchange genetic material. Antibacterial chemicals have altered the microbial populations of water, soil, and our own microbiota as a result of their addition and overuse. We are powerless in the face of antibiotic resistance. The development of new types of medications is now a requirement of the period. Bioinformatics, on the other hand, is revolutionising and intriguing the world of science and technology (Muegge, 2003). Recent advancements in affordable and quick DNA sequencing technologies have revolutionised diagnostic microbiology and microbial surveillance. To date, there are at least 47 bioinformatics resources are freely available for detecting AMR determinants in amino acids or DNA sequence data. These include CARD, SRST2, ARG-ANNOT, Genefinder, MEGARes, ARIBA, AMRFinder,

KmerResistance, and ResFinder, among others. The type of accepted input data, the presence/absence of software for searching within an AMR determinant from other resources, and use of the search approach, which can be based on mapping or alignment, are all factors that differentiate bioinformatics resources. As a result, each technique has strengths and limits in terms of AMR determinant detection sensitivity and specificity, as well as application. The listed tools can be found at public genomic data centres, downloaded from GitHub, or executed locally. Both the European Nucleotide Archive (ENA) and the National Center for Biotechnology Information (NCBI) provide online submission options for antimicrobial susceptibility sequencing and phenotypic data, allowing other researchers to dive deeper into the data and test new methodologies. Advances in whole genome sequencing, as well as the use of online technology for real-time identification of AMR determinants, are crucial in establishing control and preventative tactics to combat the growing threat of AMR. The availability of technologies and DNA sequence data is increasing, allowing for global disease surveillance also genomics-based AMR tracking. Pipelines and databases must, however, be standardised.

5.2.6 Evolutionary Studies

"Nothing in biology makes sense except in the light of evolution", stated Theodosius Dobzhansky, a brilliant American scientist. The study of evolution is essential to understanding biological challenges and improving the quality of life for humans. Kumar et al. (2008) used bioinformatics to compare genomic data from many species.

Forensic Analysis

Biomolecular data is becoming increasingly important in forensic research, and several European countries are building forensic databases to preserve DNA profiles of known offenders' crime sites and conduct DNA testing. Statistical and technological developments, such as TFT biosensors, DNA microarray sequencing, and machine learning algorithms, which give an effective manner of organising and inferring evidence, have strengthened the field (Bianchi & Lio, 2007). Nowadays, homology modelling is employed to create 3D models in order to analyse or justify our desired outcomes. Bioinformatics has changed the face of molecular research by allowing researchers to determine gene structure or sequence, protein structure, and molecular markers, as well as tie them to other structures previously known. Bioinformatics research has provided key methods for modelling a biological living cell system and docking proteins, allowing scientists to develop viable therapeutic strategies to tackle the growing problem of antibiotic resistance among infectious illness victims. Bioinformatics is concerned with the analysis and interpretation of a wide range of data, including nucleotide and protein domains, amino acid sequences, protein structures, and the behaviour of protein–ligand interactions at the molecular level. Homology modelling, often known as computational biology, is a method of evaluating biological annotations or data.

5.3 *Agriculture*

5.3.1 *Development of Drought Resistant Varieties*

Drought stress induced by unexpected precipitation is a huge danger to global food supply, and its influence is only likely to grow as climate change progresses. Understanding the impact of drought on crop and plant responses is crucial for generating superior varieties with consistent high yields to meet the growing food demand caused by a growing population relying on diminishing land and water resources. The recent introduction of innovative "-omics" technologies, like as proteomics, genomics, and metabolomics, allows us to investigate and discover genetic elements that underpin system complexity. The main challenge in this genomics era is storing and managing the large amounts of data contained in transcriptomics data or even genome scaffolds accessible for most of the plant species; it is no exaggeration to claim that bioinformatics has been well incorporated into modern-omics research. Sequence analysis and de novo genome assembly tools, similarity searching tools, genome sequencing tools, transcriptome, proteome, genome annotation tools, and metabolome analysis, as well as visualisation tools, help us analyse biological data and provide new insights into an organisation of biological systems (Dahiya & Lata, 2017). This -omics knowledge might then be applied to improve crop quality and production, as well as disease resistance and abiotic stress tolerance. Bioinformatics is changing the way molecular biology research is designed in the post-genomics age, contributing significantly to scientific knowledge while also providing new roles and perspectives to stress tolerance improvement genetic engineering programmes.

5.3.2 *Crop Improvement*

As the climate changes and the world population expands, the strain on our ability to produce enough food will increase. Fresh crop breeding and adaptation of crops to new environment required to assure continued food supply. Recent development in genomics has the potential to speed up genetically based crop plant breeding. Linking genetic data to climate-related agronomic characteristics for breeding objectives, on the other hand, remains a substantial challenge that will require the collaboration of a wide range of talents and knowledge. Bioinformatics and genomics combination has potential to help security in food wrt climate change by speeding the development of crops that are climate-ready. It makes extensive use of proteomic, metabolomic, genetic, and agronomic crop development to develop more powerful, drought-resistant, and insect-resistant crops. Cattle quality and disease resistance will improve as a result. Stress tolerance genes and alleles can be identified, which can lead to the production of stress-tolerant cultivars. Many approaches have been developed to study physiology, expression profiling, and comparative genomics. All metabolic pathways, including the glucose production process, are included in the KEGG database. Genes implicated in the ABA production pathway are heavily used in the development of drought-resistant cultivars. Researchers can use KEGG databases to find out which genes are involved in carbohydrate and ABA production. When a pathway is discovered, the genes are studied for their participation in it and in development. There has been progress in cereal varieties development that produce larger yields. These varieties will aid agriculture in flourishing in areas with inadequate soil, enabling for the expansion of the worldwide production base. Crop cultivars that can thrive in low-water conditions are being produced as well. Plant science and industry have crossed the genomics threshold with the completion of *Arabidopsis thaliana* genome sequence and the preliminary sequence for genome of the rice.

5.3.3 *Insect Resistance*

Insecticide resistance is a significant concern for insect pest control programmes in domains such as crop protection, human and animal health, and so on. The proteins encoded from a specific class of insect genes provide resistance to several pesticides. With the latest genome sequencing, high-throughput genomics, and proteomics efforts on a variety of insects, bioinformatics data processing approaches have become increasingly important to aid scientists in interpreting new information about the insects (Breton et al., 2021). As a result, bioinformatics academics and experts have begun developing specialised databases and tools for different industries. To comprehend the systems level physiology, biology, host–pathogen interaction, disease mechanisms, growth and development insect resistance, of numerous key insects, molecular biologists use high-throughput genomes, transcriptomics, regulatory genomics, epigenomics, and proteomics approaches. The massive amount of data generated by these technologies necessitates a highly logical mining and analysis of the entire data set, which may be accomplished using well-established bioinformatics approaches and tools in the field.

5.3.4 *Improve Nutritional Quality*

Nutrigenetics is the study of how food molecules, genes, and gene function interact, whereas nutrigenomics is the study of how dietary molecules, genes, and gene function interact. If nutrition researchers wish to be regarded a significant partner in the genetics and genomics arena and take full advantage of the many new opportunities, they must make a serious effort to include bioinformatics expertise into their toolboxes. Bioinformatics describes genomes, epigenomics, transcriptomics, proteomics, and metabolomics, which are all important aspects of nutrigenomics. Bioinformatics is expanding in every field of biology, and it has had a positive impact on agricultural development. By gathering, preserving, analysing, and integrating vast amounts of metabolomics, genomes, and proteomics data, bioinformatics allows users to efficiently analyse large amounts of data. In order to improve crop nutritional value and yield, bioinformatics makes data and tools available to anybody, including individuals, companies, and industries. In addition, in silico simulations can be used to detect complex interactions between protein–protein, structures of model protein, and decipher the genetic and physical high-resolution network present in plants.

5.4 *Veterinary Science*

Veterinary science study has advanced to a greater level because to bioinformatics. Bioinformatics is used in this subject to conduct sequencing research on animals such as cows, pigs, and sheep. As a result, overall productivity has increased, and cattle health has improved. Bioinformatics has also benefited scientists in the creation of new vaccine target identification methods (Pomerantsev et al., 1997).

6 Conclusion

Bioinformatics has become an important part of a variety of biological fields. Bioinformatics methods such as signal processing and image enable the extraction of conclusions that are useful from larger amounts of raw data in experimental and molecular biology. In the realm of genetics, it aids in the annotation and sequencing of genomes as well as their reported mutations. Through biological literature, text mining, and the creation of gene ontologies and biological, it aids in the organisation and querying of biological data. It can also be used to find the expression and control of proteins and genes. Bioinformatics tools help in the analysis, comparison, and interpretation of genomic and genetic data, as well as knowing the evolutionary elements in the molecular biology. It also aids in the investigation and cataloguing of biological pathways and networks on a more integrated level, that are crucial aspects of systems biology. In structural biology, it aids in the modelling and simulation of RNA, proteins, DNA, and biomolecular interactions.

References

Allaby, R. G., & Woodwark, M. (2004). Phylogenetics in the bioinformatics culture of understanding. *Comparative and Functional Genomics, 5*, 128–146.

Anderson, A. C. (2003). The process of structure-based drug design. *Chemistry & Biology, 10*, 787–797.

Arora, P. K., Kumar, M., Chauhan, A., Raghava, G. P., & Jain, R. K. (2009). OxDBase: A database of oxygenases involved in biodegradation. *BMC Research Notes, 2*, 67.

Benson, D. A, Boguski, M. S., Lipman, D. J., Ostell, J., & Ouellette, B. F. (1998). GenBank. *Nucleic Acids Research, 26*(1), 1–7.

Bianchi, L., & Lio, P. (2007). Forensic DNA and bioinformatics. *Briefings in Bioinformatics, 8*(2), 117–128.

Breton, G., Johansson, A. C. V., Sjödin, P., Schlebusch, C. M., & Jakobsson, M. (2021). Comparison of sequencing data processing pipelines and application to underrepresented African human populations. *BMC Bioinformatics, 22*(2021), 488. https://doi.org/10.1186/s12859-021-04407-x

Cantor, C. R. (1998). How will the Human Genome Project improve our quality of life? *Nature Biotechnology, 16*(3), 212–213.

Caspi, R., Altman, T., Dreher, K., Fulcher, C. A., Subhraveti, P., Keseler, I. M., Kothari, A., Kubo, A., Krummenacker, M., Latendresse, M., Mueller, L. A., Ong, Q., Paley, S., Subhraveti, P., Weaver, D. S., Weerasinghe, D., Zhang, P., & Karp, P. D. (2012). The MetaCyc database of metabolic pathways and enzymes and the BioCyc collection of pathway/genome databases. *Nucleic Acids Research, 40*(D1), D742–D753.

Cello, J., Paul, A. V., & Wimmer, E. (2002). Chemical synthesis of poliovirus cDNA: Generation of infectious virus in the absence of natural template. *Science, 297*, 1016–1018.

Chenna, R., Sugawara, H., Koike, T., Lopez, R., Gibson, T. J., Higgins, D. G., & Thompson, J. D. (2003). Multiple sequence alignment with the Clustal series of programs. *Nucleic Acids Research, 31*, 3497–3500.

Dahiya, B. L., & Lata, M. (2017). Bioinformatics impacts on medicine, microbial genome and agriculture. *Journal of Pharmacognosy and Phytochemistry., 6*(4), 1938–1942.

Eagles, N. J., Burke, E. E., Leonard, J., et al. (2021). SPEAQeasy: A scalable pipeline for expression analysis and quantification for R/bioconductor-powered RNA-seq analyses. *BMC Bioinformatics, 22*, 224. https://doi.org/10.1186/s12859-021-04142-3

Frantzi, M., Latosinska, A., & Mischak, H. (2019). Proteomics in drug development: The dawn of a new era? *Proteomics Clinical Applications, 5*, e1800087.

Gochhait, S. et al. (2021). Data Interpretation and Visualization of COVID-19 Cases using R Programming. *Informatics in Medicine Unlocked, 26*(6). Elsevier. ISSN: 0146-4116.

Greene, N. (2002). Computer systems for the prediction of toxicity: An update. *Advanced Drug Delivery Reviews, 54*(3), 417–431.

Hanash, S. (2003). Disease proteomics. *Nature, 422*, 226–232.

Imming, P., Sinning, C., & Meyer, A. (2006). Drugs, their targets and the nature and number of drug targets. *Nature Reviews Drug Discovery, 5*, 821–834.

Kumar, S., Nei, M., Dudley, J., & Tamura, K. (2008). MEGA: A biologist-centric software for evolutionary analysis of DNA and protein sequences. *Briefings in Bioinformatics, 9*(4), 299–306.

Mann, L., Seibt, K. M., Weber, B., et al. (2021). ECCsplorer: A pipeline to detect extrachromosomal circular DNA (eccDNA) from next-generation sequencing data. *BMC Bioinformatics, 23*, 40. https://doi.org/10.1186/s12859-021-04545-2

Marini, F., Ludt, A., Linke, J., & Strauch, K. (2021). GeneTonic: An R/Bioconductor package for streamlining the interpretation of RNA-seq data. *BMC Bioinformatics, 22*(610). https://doi.org/10.1186/s12859-021-04461-5

Muegge, I. (2003). Selection criteria for drug-like compounds. *Medicinal Research Reviews, 23*, 302–321.

Pomerantsev, A. P., Staritsin, N. A., Mockov, Y. V., & Marinin, L. I. (1997). Expression of cereolysine ab genes in Bacillus anthracis vaccine strain ensures protection against experimental hemolytic anthrax infection. *Vaccine, 15*, 1846–1850.

Porter, T. M., & Hajibabaei, M. (2021). Profile hidden Markov model sequence analysis can help remove putative pseudogenes from DNA barcoding and metabarcoding datasets. *BMC Bioinformatics, 22*, 256. https://doi.org/10.1186/s12859-021-04180-x

Sadraeian, M., & Molaee, Z. (2009). Bioinformatics Analyses of *Deinococcus radiodurans* in order to waste clean-up. In *environmental and computer science*, 254. Second International Conference.

Sharma, A., Ghosh, D., Divekar, N., Gore, M., Gochhait, S., & Shireshi, S. (2021). Comparing the socio-economic implications of the 1918 Spanish flu and the COVID-19 pandemic in India: A systematic review of literature. *International Social Science Journal, 71*, 23–36. https://doi.org/10.1111/issj.12266

Sinha, S. (2015). Role of bioinformatics in climate change studies. *J Science, 1*, 1–9.

Stoesser, G., Moseley, M. A., Sleep, J., McGowran, M., Garcia-Pastor, M., & Sterk, P. (1998). The EMBL nucleotide sequence database. *Nucleic Acids Research, 26*(1), 8–15.

Thompson, J. D., Higgins, D. G., & Gibson, T. J. (1994). CLUSTAL W: Improving the sensitivity of progressive multiple sequence alignment through sequence weighting, position-specific gap penalties and weight matrix choice. *Nucleic Acids Research, 22*, 4673–4680.

Zhang, L., & Hong, H. (2015). Genomic discoveries and personalized medicine in neurological diseases. *Pharmaceutics, 7*, 542–553.

Index

S. Dutta and S. Gochhait (eds.), *Information Retrieval in Bioinformatics*,
https://doi.org/10.1007/978-981-19-6506-7